Exploring Genome's Junkyard:
In the labyrinth of evolution

Subir Ranjan Kundu[1] (Author)

Arkatapa Basu (Managing Editor)

[1] Corresponding Author: subirranjankundu@gmail.com

Exploring Genome's Junkyard:
In the labyrinth of evolution

Subir Ranjan Kundu (Author)

Arkatapa Basu (Managing Editor)

Academica Press
Washington~London

Library of Congress Cataloging-in-Publication Data

Names: Kundu, Subir Ranjan (author)
Title: Exploring genome's junkyard : in the labyrinth of evolution | Subir Ranjan
Kundu
Description: Washington : Academica Press, 2022. | Includes references.
Identifiers: LCCN 2022944867 | ISBN 9781680538687 (hardcover) |
9781680538700 (paperback) | 9781680538694 (e-book)

Dedication

To my friend Merelin Keka Adhikari, (Capacity Building & Training Advisor, Urban Sanitation Programme, Dhaka, Bangladesh) of WWW.SNV.ORG, Netherlands. The continuous flow of inspiration, instantaneous wit, and pleasant empathy of her helped me a lot in boosting up my enthusiasm to carry on my journey of life through the unavoidable, monotonous and secluded state of life, unless writing of this book would have remain incomplete.

Contents

Preface

The modern world is bombarding us with a huge amount of information, the reliability of which is often questionable. We chase sensation, often having more faith in science-fiction, while real science seems boring and, above all, incomprehensible. Therefore, we need guides who will lead us through the complexities of knowledge, explain in an accessible way not only what is happening around us, but, most importantly, also what we are part of. And even though it is basic knowledge, necessary to answer not only the question 'Where are we from?' but also 'Where are we going?,' it is becoming less and less understandable and accessible. It also seems to be distant from everyday life and practically irrelevant. So, if we do not want to completely separate the world of science from life, not to be fooled by false theories but to seek the truth, we need non-fiction books.

Dr. Subir Ranjan Kundu in his book "*Exploring Genome's Junkyard: In The Labyrinth of Evolution*," once again takes us on a fascinating journey through the nooks and crannies of evolution. The author skillfully navigates difficult fields such as molecular genetics, genomics and evolutionary science. The book undertakes the task of critically reviewing ideas, hypotheses and evidence about biological evolution, particularly with regard to the human genome. It presents the modern discoveries of DNA analysis and the molecular traces of mutational changes, referring primarily to the human species. We can read the critical research observations of molecular geneticists, evolutionary biologist and genomicists who claimed all humans are common offsprings, genetically related to a primordial couple (mitochondrial Eve and Y-chromosomal Adam). The author traces molecular pathways of origin, evolution and diversification of the anatomically modern human species. This unique story of molecular introspection of anthropogenic evolution can be traced back to the emergence of the unicellular, prokaryotic organisms in water, which supposedly emerged on Earth around 4 billion years ago. We learn about the role of DNA in the progress of biological evolution and its molecular footprints of mutational changes which control the evolutionary changes in a simple way. The proposed book attempts to revise the conventional view of biological evolution as a two-step process mutation followed by natural selection. The work focuses on the author's proposal of the origin and evolutionary developmental transitions of biological existence in general, and of the human genome, in particular, on Earth, in the context of a variety of ideas that can be fitted into a separate set of evidence. We start with the conceptual gene introspection and gradually move towards the modern understanding of the structural making of the human genome, its functional role that makes the biological entity ascribed to fit evolution occupying the epicentre of the principles of biological evolution. In this unique literature, Dr. Subir Ranjan Kundu has tried to focus on this evidence to expose the concept of non-coding DNA repetitive elements and their effect on higher organisms. Though, most molecular geneticists and genomicists have promulgated that the majority of DNA is biochemically vigorous and genetically functional, reality is quite different. One of the major unsolved mysteries in human genetics is the proportion of functional and non-functional sequences in the genome. According to the contemporary concept of evolution, there is a sharp difference between biochemical activities and functionalities. Such critical observations of molecular geneticists and genome biologists have helped us to reach a straightforward conclusion that the existence of excess DNA has no role in evolution as it is junk. In reality, non-coding sequences of DNAs are not junk material that is attached to the coding sequence, but essential, as it plays a cardinal role as the potential hotspot for the mutations which is the ultimate driving force in the evolution in the biological world. This book provides a scientific narrative regarding the structural and functional features of DNA, in particular 'Dark DNA,' and aims to show the identity of hidden genes (commonly thought of as missing genes) that have gone through a series of mutational cycles to steer the molecular evolution of a biological entity on Earth. The author has outlined a very complex picture, combining a unique synthesis of modern discoveries in DNA sequencing with critical observations on mutagenesis in molecular genetics, gathering analyses from phylogenomics and unique discoveries of molecular genetics.

The book is aimed at advanced science enthusiasts, researchers, PhD students and students of Evolutionary Biology and Zoology who want to enrich their knowledge in the field of molecular and evolutionary biology, particularly, to learn about the new concept of non-coding genome, the contemporary facts and findings on so-called molecular mutation of missing genes hidden in Dark DNA in terms of its biochemical activities and functionalities.

Dr Rafał Kurczewski
Deputy Director
National Park of Wielkopolska
Poznan University of Life Sciences
JEZIORY 62-050 MOSINA
POLAND
Tel 61 89 82 304
Fax 61 89 82 301
Kom. 500 401 084

Acknowledgments

I express my deep sense of gratitude and indebtedness to Dr. Parth Pratim Dhar, Principal/Professor, Sir Ashutosh Govt. College, Chittagong, Bangladesh; Dr. Nikunj Bhatt, Associate Professor in Zoology, Department of Biology, Vittalbhai Patel & Rajratna P T Patel Science College, Gujarat, India, and Dr. Rafal Kurczewski, Deputy Director, National Park of Wielkopolska, Mosina, Poland; for their suggestions and guidance in this present writing.

I am especially pleased to Dr. S. Sundar, Research Scientist, S.S. Research Foundation, Kallidaikurichi, Tamil Nadu, India; who has advised me in drafting the different part of the manuscript, required for further reviewing consultations of literature, library consultation and seeking opinion of the subject matter experts etc.

I am also thankful to Susan Zadek and Sung-Min Han, Albert Campbell Library; Gail MacFayden and Eva Lew, Bloor Gladstone Library, under Toronto Public Library Network, Toronto, Canada for extending Library facilities and providing necessary help in consulting the literature required for this project to undertake.

I am expressing my gratitude to thank all authors, editors, publishers of books and journals, colleagues for sharing the printed and digital materials, providing valuable suggestions and tendering necessary clarifications during the last few years, selflessly helped me out to complete this task.

Introduction

"Mendelian Factor" to "Johannsen's Gene": Evolutionary Journey of "Nuclein" Through Temporal Cascade

I am convinced that it will not be long before the whole world acknowledges the results of my work.

---- ***Gregor Johann Mendel***

The journey of chromosomal study started with a miraculous hybridization experiment of sweet pea (*Pisum sativum*) plants in a garden adjacent to the Augustinian monastery of St. Thomas or Königskloster at Brunn, Austria, where a humble pastor and the "father of genetics," one of the greatest naturalists, Gregor Johann Mendel, lived in 18th century. Mendel started his experiment in 1857, critically observed and statistically analyzed the transmission of phenotypic or visible characteristics of pea plants from parent plants to its offsprings. After nine years of experimental trials and mathematical analysis, Mendel's flawless, elegant and simple observations appeared in the annual proceeding of the Natural History of Brunn in 1866. The constant transfer of visible characteristics from parent to offspring in a certain pattern laid the stepping stone for the maiden principles of heredity, popularly recognized as the laws of hereditary/principles of Mendelism. It was an amazing discovery, and a milestone in the history of molecular biology that mesmerized the entire science community and millions of followers of life science and biological science. The mesmerizing factor of the Mendel's law of heredity was questioned as it was recognized as two complementary laws. The two laws comprised independent assortment and laws of segregation which was a simplistic formulation of transmission of hereditary principles passed on from generation to generation in a certain pattern by expressing visible (phenotypic) characteristics.

Mendel's eloquent analytical narration of hybridization experiment on the sweet pea successfully portrayed the biological relationship of evolutionary inheritance from the parent to its offspring in the plant kingdom which was found to be applicable in animal kingdom too. Mendel's experimental analysis clearly elucidated that recombination of parental traits (two distinct characteristics of same trait, coming from two distinct parental lineages), were witnessed in the offspring. Hence, it was clear to him that segregation of the factors must be effectively completed prior to the start of recombination to yield a unique (de novo) visible feature (phenotypic expression).

Mendel's invisible component is the true determining "factor," which controls the principles of biological inheritance from predecessor to successor and continue the journey of evolution. Nowadays, a high school student is aware that the determining "factor" which regulates the principles of inheritance from the ancestor to its successor is nothing but gene. The calm, eloquent and enigmatic pastor/academician, Gregor Johann Mendel was optimistic that his observation would be acknowledged and appreciated by the world of science. It would take his findings to the next level but that did not happen. From 1866 to 1900 was a phase of hibernation when Mendel's work had plunged into obscurity, the world of science hardly recognized Mendel's work as fundamental research work anymore, however Dutch botanist Hugo de Vries, German botanist and geneticist Carl Erich Correns and Austrian botanist Erich Tschermak von Seysenegg rediscovered Mendel's work on the principles of heredity. Their findings corroborated Mendel's principles of biological inheritance. Instead of pursuing and patronizing their own principles of heredity, these three scientists with noble and selfless spirit upheld the principles of inheritance, once propounded by Gregor Johann Mendel, which was then recognized as "Mendel's Laws of Inheritance."

At the beginning of 19th century, when cellular and ultracellular structures were being discovered, it was widely accepted that the structural and inherent entity of the hereditary factor of germ cells which passed from generation to generation, maintained and regulated the visible characteristics of the offspring. So, cellular and ultracellular study of any living entity helped to decipher the inherent hereditary factor or profile (genotypes), which regulated the phenotypic expression of the concerned organism. Whereas, the lack of basic findings of cellular and ultracellular structural and functional aspect of any living organism in 18th century was a real challenge for pastor Mendel as structural identity and functional cell, specifically of germ cell and concepts of genotype were not present. Naturally, it was difficult to describe the effect without knowing the cause that regulated the changes which Mendel recognized as "factors." Unfortunately, Mendel's simplistic mathematical as well as statistical observation on biological principles of heredity and biological evolution was either considered unconvincing or was ignored by naturalists, scientists and academicians of biological science of that time. So, the statistical approach to the solution of heredity remained unexplored.

Though the journey of the scientific investigation of biological evolution in relation to cytogenetical as well as chromosomal studies started with mathematical/statistical analysis rather than technology-guided microscopic and ultramicroscopic trails, the discovery itself was credited to Johnanes Friedrich Miescher, an acclaimed Swiss medical physician and biochemist, who worked under the supervision of Hoppe-Seyler of Tubingen, a renowned biochemist. He successfully extracted and identified the hereditary material known as "nuclein." Miescher's discovery of nuclein was purely a matter of chance as he was investigating a protein extracted from leucocyte cells. When he noticed that the substance he extracted did not biochemically resemble any protein, he named it "nuclein" in 1869 (Dahm, 2005). The evolutionary journey of DNA (Deoxyribose Nucleic Acid) started with the birth of "nuclein," its predecessor in the 18th century. Ideally, if Mendel and Miescher could work together or if their work overlapped to yield synergistic result, where Miescher's structural narration of "nuclein" defined the functional trail of Mendel's "factor," interpretation of biological evolution would have occurred earlier. However, that did not happen. It took a long way to traverse through the trail of genetics and molecular biology to arrive in the era of Darwinian natural selection of biological evolution in the 18th century, followed by Myer's molecular evolution era in the 21st century.

Every story has a preface, likewise, every experiment has its early episodes. The discovery of cell, nucleus and nucleic material started with the invention of microscope in 1674 by Antonie Philips Von Leeuwenhoek followed by his next publication "Letter on the Protozoa" in 1677. It took more than 150 years for a Scottish botanist, who studied the "areola" of orchids under a microscope, to recognize those as nucleus of a cell in 1831(Lane, 2015). Though, the basic structure of the cell was discovered by

Robert Hooke, an eminent English natural philosopher and poly-math, in 1665, the recognition of structural and functional entity remained a matter of dispute in the world of biological science for more than a century. Old-school scientists and natural philosophers expressed their skepticism about its identity as the universal unit of life and its regeneration process from time to time. Finally, in 1839, Matthias Jakob Schleiden and Theodor Schwann propounded the "Cell Theory," which clearly established the structural identity, physiological and hereditary entity of the cell. It further established that being the key constituent of the cell, nucleus intermittently transfers from generation to generation and plays an important role in the different phases of cell division (Schleiden, 1839; Schwann, 1839). In 1842, Carl Wilhelm Von Nageli, a Swiss botanist, was one of the staunch critics of hereditary principles of Mendel and Darwinian doctrine of natural selection. He was conducting a researcher on pollination and cell division (meiotic cell division) of different plants, and identified subcellular structure, later recognized as chromosomes, though his discovery of "Idioplasma" and interpretation of "interlinked meshwork of string-like structures throughout the bodies" of the concerned organism was proved wrong. Wilhelm Hofmeister (1848), a renowned biologist, noticed the thread-like bodies in the nucleus for the first time and discovered chromosome, the key nuclear structures which act as hereditary material of living cells. Freidrich Anton Schneider, a German zoologist/taxonomist, experimentally demonstrated that the intermittent transfer of germplasm (nucleus) in the cell divisional phase and the cytogenetical changes in the nuclear material like its condensation and movement during divisional phase of flatworm eggs belonged to Platyhelminthes *(Mesostomum ehrenbergii)*. The successful fusion of sperm nucleus and egg nucleus of sea urchins by an embryologist, Wilhelm August Oscar Hertig in 1876, inspired the great Belgian scientist, Edouard Van Beneden to seek better insight into the history of the nuclei. He also worked on the model organism *Ascaris megalocephala* in which he noticed that the chromosomes of the oocyte nucleus failed to fuse with the chromosome of male nucleus, which made him conclude those nuclei as distinct cyto-hereditary entities (Brind'Amour and Garcia, 2007; Hamoir, 1992). The successful completion of Van Beneden's cytological studies in 1883 helped the world of biological science to reach the zenith of cytogenetical discoveries. His experimental demonstration was able to prove that chromosomes of *Ascaris megalocephalis*'s offspring, received from their parents, were in equal proportion and were transferred through the union of nuclei of their respective germ cells. It was the first successful attempt at meiotic cell division (Hamoir, 1992). Eduard Adolf Strasburger (1877, 1884), a Polish-German botanist, made identical observation in plants by studying the cell division pattern of different plants (e.g. Orchids) to elucidate the role of nucleus and chromosome in relation to heredity.

In 1879, Walter Flemming, an eminent German anatomist, who is also recognized as the "founder of the science of cytogenetics" meticulously studied the chromosomal behavior in the cell nucleus during somatic cell division (Mitotic cell division). In his experimental observations, Flemming used aniline dye for the first time to reveal the well-stained, rod-like shapes of the nucleic material (under microscopic field), recognized as "chromosome," which he wrote about in his book "Zell-substanz, kern und Zellthielung"[2] (Flemming, 1882). Though neither Van Beneden nor Flemming used the term "chromosome," it was the German cytogeneticist, Wilhelm Waldeyer, who in his publication "Uber Karyokinese und ihre Beziehungen zu den Befruchtungevorgiingen" coined the term for the first time in 1888 (Cremer and Cremer, 1978). It needs to be mentioned that Flemming specifically observed that the division of chromosomes in somatic cell was found to be numerically equal in parental and newly-formed cells, so this type of divisions was recognized as mitotic or equal division. Whereas, division of chromosomes in germ cell yielded reduced number of chromosomes in newly formed cells with respect to parental lineages, hence, it was recognized as meiotic or reduction division.

Further along in the cytogenetical study, Wilhelm Roux (1883), a German zoologist and a pioneer researcher in experimental embryology, revealed that cell division not only divided or replicated and transferred nuclear material from cell to cell or from parent to offspring but also transferred its inherent qualities from parent to offspring, which act as hereditary regulator of any trait.

In the history of chromosomal studies, German biologist Hermann Henking was the first to discover the identity of sex chromosome or allosomes, which play an important role in sex determination of biological organisms. During his study of mitotic cell division pattern of wasp cells, Henking observed observed that some wasp cells have 11 chromosomes,

---

[2] Zell-substanz, kern und Zellthielung (In German): Cell-substance, Nucleus and Cell-division (In English).

while others have 12 chromosomes. During the meiotic division of sperm cells, Henking noticed that the 12th chromosome looked different than rest of the 11 chromosomes. He could not find the same chromosome in the egg of the female wasp, which helped Henking determine that the 12th chromosome is the sex determine chromosome or "X chromosome" in his hypothesis, which he promulgated in 1891 (Brown, 2003).

In 1893, August Freidrich Wesimann, an evolutionary biologist and eminent professor from the University of Freiburg, Germany, propounded the "Germplasm theory" or heredity, in which he suggested that the hereditary material was contained in the nucleus of germ cell, and it was recognized as "germplasm."

In 1901, Clarence E. McClung, a zoologist and cytologist, successfully established the regulation of chromosomal action through the inheritance of certain characters in Orthoptera. While studying its spermatogenesis, he came across the heterochromatic chromosome found in the male individuals which could not be found in females. He identified these as "accessory chromosomes" and concluded that sperms, containing the heterochromatic chromosomes, are sex determinants and are responsible for producing male offsprings (McClung, 1901; Brown, 2003). In 1902, two great researchers in the field of cytogenetics-Walter Sutton from Columbia University, USA, and Theodor Boveri, from University of Urzburg, Germany, propounded "Boveri-Sutton Chrosome theory." They observed that functional pattern of chromosomes were in tandem with the Mendelian principles of inheritance. They hypothesized that chromosome contained inheritable factors or genes that passed down from ancestor to offspring (Sutton, 1903; Boveri, 1904).

The history of structural and functional entity of gene started with tracing out its identity and mode of function, by diagnosing the chromosomal structure and functions, initiating and practicing under add-up modifications of an important histochemical test. For over two centuries, it was popularly known as Feulgen cyto-histochemical reaction, which had been used widely in the field of applied biology and medicine (mainly in clinical diagnostics) since the beginning of 19th century and continuously used in applied genetics and molecular biology. The Feulgen nuclear reaction technique was introduced by an eminent cytogeneticist, Robert Feulgen, and his co-worker, Heinrich Rossenbeck, in 1924 to stain the histochemical preparations collected from different biological organisms *in-*

situ. They applied the erstwhile biochemical principles of detecting organic aldehydes found in the biological tissues, determined by colored reactions which was devised by Hugo Schiff in 1866 (Feuelgen and Rossenbeck, 1924). Feuelgen and Rossenbeck had successfully standardized a biochemical protocol called HCL hydrolysis which yielded "free aldehydes in the DNA backbone structure." It further responded to the colored reaction of Schiff's Test (Chieco and Derenzini, 1999).

In the history of chromosome research or nucleic acid research or structural and functional aspects of DNA, Feulgen Reaction was an impetus as well as a technological shift from cytogenetical investigation to cytochemical investigation in DNA studies. Nowadays, even a student in high-school is familiar with the terminology "gene" which gives expression to heredity. "Gene" was found to be responsible for expression of characteristics in organisms. This term was coined by the Danish biologist, Wilhelm Johanssen in 1909.

The term "factor" used by Mendel in 1860 was found to be the key determining factor of the biological organism which regulated the transfer of phenotypic characteristics of the organism. It is inherited by the offspring from the parent and is biologically synonymous to "gene." Johannsen, in the 19th Century, said that "any particle to which the properties of a Mendelian factor may be attributed is called gene" in his book "Elemente der exakten Erblichkeitslehre," published in 1909 (Johanssen, 1909; Anonymous, 1911). Strategically, genes are found in chromosomal threads which split apart and recombine during the process of meiotic or reduction division of germ cells (possessing haploid set of chromosome) or gametes (n). Somatic number or diploid set of chromosomes are restored as a result of fusion of maternal and paternal nuclei (composed of nuclein or a compound formed of nucleic acid and protein) to form the zygote of diploid set of chromosomes (2n).

Following the footsteps of McClung (1901), Thomas Hunt Morgan, a renowned academician of Columbia University, studied the sex-linked inheritance of drosophila (fruit flies) through breeding experiments in 1910. After careful analysis of a large number of fruit flies with red eyes he found an individual with white eyes. As Morgan engaged in breeding the white-eyed, mutant fly, he noticed that white-colored eyes were linked to male offsprings. This helped Morgan understand that eye color and sex determining factor are attributed to the same chromosome. Morgan further concluded that genes are contained in chromosomes and these genes regulate inheritable traits which is transferred from predecessor to successor (Gleason, 18th May, 2017).

Coming a long way from 1888 to 1926, the isolated efforts of Johanssen, Boveri, Sutton, McClung, Morgan and other geneticists and cytogeneticists paved the way for the discovery for the Mendelian hereditary factors which were later identified as the linearly-arranged genes on chromosomes. The earnest search for the functional aspects of genetic inheritance by these scientists reached the molecular level. The structural formation of "gene" which is responsible for regulating the principles of inheritance, was yet to be explored. In 1928, Fredrick Griffith, one of the renowned British bacteriologist and epidemiologist, witnessed a unique instance of genetical transformation in bacteria when he experimented with two different strains of *Pneumococcus* (*Streptococcus pneumoniae*)—S-strain, characterized by shiny colonies (having polysaccharide, mucous coating) and R-strain characterised by rough colonies, (having no polysaccharide coating). In laboratory conditions, Griffith noticed that the mice infected with S-strain (virulent) died of bacterial infection whereas, the ones with R-strain infection (avirulent) did not die. In his experiment, Griffith heat-killed the S-strain and R-strain of Pneumococcus and injected it into the first set of mice which remained alive. However, he noticed that a mixture of heat-killed S-strains and live R-strains Pneumococcus killed all the mice while living S-bacterial strain could be recovered from them. This helped Griffith to understand that the "transforming principle" of the dead pneumococcus S-strain successfully transformed the R-strain to S-strain (Griffith, 1928; O'Connor, 2008). Hence, the successful transfer of genetic material which enables to transform the cytogenetical entity of any organism from one form to another was established by Griffith (1928). However, biochemical as well as structural identity of gene as hereditary material was yet to be discovered.

A promising German cytogeneticist, Emil Heitz, strived to carry forward this legacy of his predecessors. This young scientist successfully developed a unique but inexpensive staining technique known as "Kochmethode" to prepare the chromosomes in-situ, collected from Moss samples, which was followed by the next level of application on chromosomal preparation collected from 115 plant species, fruit flies and a number of diptera. This staining technique helped him to recognize "cytologically detectable longitudinal differentiation of chromosomes" in relation to the linear arrangement of genes in it (Passarge, 1979). After continuously examining his experimental results from 1928 to 1935, Heitz finally revealed the longitudinal differentiation of chromosomes in two distinct entities of chromosome—heterochromatin (which remains condensed and uncoiled in the different stages of cell division starting from prophase to telophase), genetically inert areas and euchromatin (the lightly stained location which is genetically active and start unwinding during telophase). Heitz also discovered a giant chromosome known as salivary gland chromosome of diptera and formation of satellite chromosomes integrated in nucleolus (Passarge, 1979).

Following in the footsteps of Fredrick Griffith, Oswald Theodore Avery, a Canadian-American medical researcher and molecular biologist along with two fellow scientists, Collin Munro Macleod, a geneticist and Maclyn McCarty, American geneticist, endeavored to trace the structural as well as biochemical identity of the "transforming principle" of Fredrick Griffith's genetical recombination experiments between 1933 and 1944. Though, it was revealed earlier that "transforming principles" or genes are made up of DNA and proteins, the protein was considered genetic material and DNA was structurally and functionally found to be less variable than protein (O'Connor, 2008). Avery and his colleagues isolated the biochemical molecules—DNA, RNA, and proteins— from the heat-killed S-strains of pneumococcus to identify whether its protein or nucleic acids have such "transforming principle" to change R-strain to S-strain in the pneumococcus (Avery, MacLeod and McCarty, 1944). Avery and his colleagues pre-treated the heat-killed S-strain cells with DNase, RNase and Protease enzymes with permutation-combination method and noticed that application of protein digesting enzyme (Protease) and RNA digesting enzyme (RNase) did not affect the transformation process. Whereas, the application of DNase stopped the bacterial transformation. This helped Avery and his colleagues conclude that DNA is the ultimate biomolecule which triggers the genetic transformation (O'Connor, 2008). Works of Avery and his associates convinced the world of science and evolutionary biology, in general, to accept that DNA is that genetic material which was previously called the Mendelian herediary factor by earlier researchers of the 18th century.

In his eloquent literature on cytogenetics, "Evolution and Genetics," Thomas Hunt Morgan, clearly defined "any particle to which the properties of a Mendelian factor may be attributed is called gene" (Morgan, 1925).

Following the footsteps of Morgan, in 1935, Hermann Joseph Muller, the American geneticist, along with geneticist A.A. Prokofyeva-Belgovskaya from Russia (erstwhile USSR), focussed on the chromosomes in the salivary glands of fruit flies, particularly on the "inert region"[3] of X-chromosome. The Y-chromosome was found to contain few discs at frequent intervals, having few genes. He stated that "a minimal particle of the chromosome which can be separated from other particles by X-ray breakage" (Muller and Prokofyeva, 1935). In 1939, two eminent geneticists, George Wells Beadle and Alfred Henry Sturtevant (1939), portrayed the determining properties of gene which helped to distinguish an animal from a plant or a species of plant from another at a taxonomical level. The empirical judgement of the futuristic evolutionary mode regarding whether a certain organism is going to develop superior structure and better functioning or the derived population would be in the regressive mode of evolution. So, in the field of cytogenetical studies, the endeavor of the cytogeneticists in search of principles of inheritance, which started with the discovery of nuclein concluded with the recognition of the Mendelian Factors, as genes in the middle of the 19th century.

The ultimate affirmation of the identity of DNA as genetic material was further confirmed with an experiment by Alfred Day Hershey, the renowned American bacteriologist and geneticist, and Martha Cowles Chase, the great American geneticist. Their experimental work on bacteria-infected virus, or bacteriophage, confirmed the identity of DNA as genes in 1952 (O'Conner, 2008). The experiment showed that the bacteriophage attached itself to the bacterial wall of its host and injected its genetic material in to the bacterial cell. The host bacterial cell, in turn, incorporated the viral genetic material. It was a strategic advantage for Hershey and Chase to discover whether the genetical material was protein or DNA in nature. In order to find out, they grew a batch (batch 1) of phage culture in culture medium, prepared with radioactive phosphorous. Another batch (batch 2) of phage virus was grown on the culture medium prepared with radioactive sulfur so that the new-born phages from batch 1 have radioactive DNA (as radioactive phosphorus was incorporated in the DNA molecule of new phages) and the newborn phages from batch 2, have radioactive protein (as radioactive sulfur was incorporated in the protein of new phages) (O'Connor, 2008). After

---

[3] Inert Regions: Most of the chromatin fibrils in this region is found to be genetically inert during mitotic division, so it has been assumed having less number of genes.

completion of the first step, Hershey and Chase (1952) infected the bacterial culture of *Escherichia coli* with both of the newly born phages so as to transfer the viral genetic material inside the *E.coli* and to incorporate or blend the viral genetic material with the bacterial genome. Herchey and Chase were anticipating three possibilities (Hernandez, 23rd June 2019)-

- *a.* Presence of radioactive phosphorous and radioactive sulfur in the blended genetical material isolated from the infected bacterial cells of *E. coli.*

- *b.* Presence of radioactive sulfur but absence of radioactive phosphorus in the blended genetical material isolated from the infected bacterial cells of *E. coli.*

- *c.* Presence of radioactive phosphorus but absence of radioactive sulfur in the blended genetical material isolated from the infected bacterial cell of *E. coli.*

However, careful review of the experimental results of the transmission of genetical material from bacteriophage to *E-coli* found the last possibility to be successful. It further confirmed that as an integral part of chromosome, it was not protein but DNA that acted as hereditary as well as genetic material. So, the compilation of isolated facts and findings of cytogeneticists since the middle of the 18th century conveniently defined gene as-

- a. The ultimate unit of chromosome which remains undivided, functionally and structurally, and that any chromosomal aberrations or crossing over does not change its functional and structural entity.

- b. The unit of heredity, transferred from parent to offspring.

- c. The unit of self-reproduction.

- d. The ultimate unit that recombines unique characteristic in offspring, inherited from parents.

- e. The unit of mutation or permanent alteration of the herediary material or gene of a living organism at a cellular level.

- f. The ultimate functional unit of living organisms that coordinates different functions at the cellular level and regulates the biochemical and metabolic reactions at the molecular level.

Finally, the invention of 3 dimensional double helix model of DNA, which was theorized in 1952 by James Dewey Watson, the eminent American molecular biologist/zoologist, and further supplemented by the works on DNA replication, transcription and protein translation process, that revealed the structural formation of gene and gene action, done by Francis Harry and Compton Crick from 1950-1970, the famous British molecular biologist/neuroscientist. It led to a shift in paradigm which helped geneticists to transition from an era of ctyochemical studies of genetics to an era of molecular biology (Ebach and Holdrege, 2005). Since 1953, Watson-Creek's modelling and interpretation of DNA, the ultimate biochemical component of chromosomes, helped the world of science to acknowledge gene as the functional unit of the DNA molecule.

However, the biochemical visualization or portrayal of a chromosome presented it as a complex biomolecule made of a number of simple biomolecules: primarily comprising two distinct types of nucleic acid (and RNA), histone and non-histone proteins.

Further biochemical investigation of the chromosomal ultrastructure and functions was carried forward by the advancement of technologies, development of cytochemical analyzing protocols, tireless effort and endeavor of biochemists, geneticists, molecular biologists through a series of roadblocks in the last three centuries.

The earliest endeavors of the cytogeneticists to coordinate structural analysis of chromosomes studies revealed some distinct differentiation between the chromosomes of a prokaryote (e.g. bacteria, unicellular algae, virus etc.) and eukaryote (e.g. flowering plants, birds, mammals etc.). The structural formation of a simple chromosome, which is also recognized as genophore, is mainly composed of a long, linear array of DNA molecules (though nucleic material in some RNA virus or DNA-RNA virus contain RNA or DNA-RNA strands). On the other hand, more complex nature of eukaryotic chromosome is composed of two different types of nucleic acid like DNA, RNA, proteins histone and non-histone type. Even if the structural and functional aspects of DNA in a prokaryote is compared to its eukaryote counterpart, some distinct features are noticed-

a. Predominantly, prokaryotes possess haploid (n) set of chromosome, whereas, eukaryotes possess diploid set of chromosome (2n).

b. DNA of prokaryotes are found to be scattered in the cytoplasm (everywhere without adherence to any cell organelle). While, the DNA of the eukaryotes are confined in the nucleus, a distinct organelle in the cytoplasm.

c. DNA of prokaryotes possess round-shaped plasmids which are structurally not called chromosomes but functionally act like chromosomes, whereas, DNA of eukaryotes are found to possess linear chromosomes.

d. Replication of DNA in prokaryotes are not attributed to any organelle, whereas, DNA replication of eukaryotes are found to be attributed to the centrioles.

However, the cytochemical analysis along with ultramicroscopic investigation, high resolution auto-radiographic analysis, micro-spectrometric analysis etc. of chromosomes, nucleic acids, proteins and its structural units, amino acids revealed the ultrastructural and functional aspects of chromosome in general and gene action in particular. It was noticed by the cytogeneticists and molecular biologists that cytochemical analyzing tools like Feulgen nuclear reactions helped them to analyze qualitative estimation of aldehydes which are integrated in the backbone of DNA. Whereas, the Dische reaction, where diphenylamine was used to develop a color reaction, helped to analyze deoxy sugars (Dische, 1955). For precise estimation of DNA and RNA, "Flow cytometric estimation" was devised by Mack Fulwyler, on the basis of Coulter principle which was invented by Wallace Coulter in 1953 (Fulwyler, 1965). In 1980, the enzyme digestion technique to extract nucleic acid was devised by Arun Sharma and Archana Sharma. They used nucleases to catalyze deoxyribose sugar in DNA and ribose sugar in RNA. They also used enzyme pepsin to catalyze histone proteins and chymotrypsin to catalyze non-histone proteins (Sharma and Sharma, 1980). To detect the nitrogen base pairs (purines and pyrimidines) in the DNA sample, ultraviolet light absorption spectroscopic technique was introduced at the 2650 Angstrom wave length (Sharma and Sharma, 1980). To investigate the chromosomal proteins further (e.g. detection of non-histone chromosomal protein, immunofluorescent assay like immunofluorescent staining of Sodium Dodecyl Sulfate polyacrylamide gel electrophoresis was used since late nineties, which was invented by Ulrich Laemmli in 1970), a series of advanced cytochemical and histochemical, chromosomal techniques were used in the last two centuries. So, in order to understand the structural formation and functional aspects of chromosome, DNA and gene, a preliminary ideas on the molecular footprints of inheritance is a pre-requisite to traverse the path of nucleic acid and explore the world of DNA.

References

Avery, O. T., Macleod, C. M., & McCarty, M. (1944). Studies on the chemical nature of the substance inducing transformation of pneumococcal types: induction of transformation by a desoxyribo-nucleic acid fraction isolated from pneumococcus type iii. *The Journal of experimental medicine*, *79*(2): 137–158.
https://doi.org/10.1084/jem.79.2.137

Anonymous (1911). "Professor Johannsen's Columbia Lectures." Science. 34 (876): 484. doi:10. 1126/science.34.876.484

Axelsson, E., Webster, M. T., Smith, N. G., Burt, D. W., and Ellegren, H. (2005). Comparison of the chicken and turkey genomes reveals a higher rate of nucleotide divergence on microchromosomes than macrochromosomes. *Genome research* 15(1): 120–125.
https://doi.Org/10.1101/gr.3021305

Beadle, G. W. and Sturtevant, A.H. (1939). An Introduction to Modern Genetics.New York: The Macmillan Company.
https://archive. org/details/ introductiontomo00wadd (Accessed March 9, 2017).

Boveri, T.H. (1904). Ergebnisse über die Konstitution der chromatischen Substanz des Zelkerns. Fisher, Jena.

Brind'Amour, K., and Garcia, B. (2007). "Karl Wilhelm Theodor Richard von Hertwig (1850-1937)." Embryo Project Encyclopedia. ISSN: 1940-5030
http://embryo.asu.edu/handle/10776/1706.

Brown, S. (2003). Entomological contributions to genetics: Studies on insect germ cells linked genes to chromosomes and chromosomes to Mendelian inheritance. *Archives of Insect Biochemistry and Physiology* 53: 115–118.

Burt, R. S. (2002). "The Social Capital of Structural Holes." Pp. 148–92 in The New Economic Sociology, edited by Mauro F. Guille´n, Randall Collins, Paula England, and Marshall Meyer. New York: Russell Sage Foundation.

Chieco, P. and Derenzini, M. (1999). The Feulgen reaction 75 years on. Histochem Cell Biol 111: 345-358.

Cremer, C., and Cremer, T. (1978). Considerations on a laser-scanning-microscope with high resolution and depth of field. Microscopica Acta 81:31–44.

Crick, F.H.C. (1958). On protein synthesis. *Symp. Soc. Exp. Biol.* 12: 138–1163

Dahm, R.(2005). Friedrich Miescher and the discovery of DNA. Developmental Biology 278(2): 274-288.

Dische, Z. (1955). In: The Nuceic Acids, ed. by Chargaff, E. & Davidson, J. N. New York: Academic Press Inc.

Ebach, M. C., and Holdrege, C. (2005). DNA barcoding is no substitute for taxonomy. *Nature* 434(7034):697. doi:10.1038/434697b

Feulgen, R. and Rossenbeck, H. (1924). Mikroskopisch-chemischerNachweis einer Nukleinsàure von Typus der Thymonuklein-sàure und die darauf beruhende selektive Fàrbung vonZellkernen in mikroskopischen Pràparaten. Hoppe-Seylers Z Physiol Chem 135:203-248

Fillon, V. (1998). The chicken as a model to study microchromosomes in birds: a review. *Genet Sel Evol* 30: 209.
https://doi.org/10.1186/1297-9686-30-3-209

Flemming, W. (1882). Zellsubstanz, Kern und Zelltheilung (F. C. W. Vogel, Leipzig).

Fulwyler, M.J. (1965). Electronic separation of biological cells by volume. Science.150:910–911. doi: 10.1126/science.150.3698.910.

Gleason, K. (18th May, 2017). "Calvin Bridges' Experiments on Nondisjunction as Evidence for the Chromosome Theory of Heredity (1913-1916)." Embryo Project Encyclopedia. ISSN: 1940-5030
http://embryo. asu.edu /handle/10776/11500.

Griffith, F. *(1928).* "The Significance of Pneumococcal Types." *Journal of Hygiene. Cambridge University Press.27 (2): 113–159.*
doi: 10.1017/S0022172400031879

Hernandez, V. (23rd June, 2019). "The Hershey-Chase Experiments (1952), by Alfred Hershey and Martha Chase." Embryo Project Encyclopedia. ISSN: 1940-5030
http://embryo.asu.edu/handle/10776/13109.

Hamoir, G. (1992). The discovery of meiosis by E. Van Beneden, a breakthrough in the morphological phase of heredity. *Int. J. Dev. Biol.* 36: 9-15

Hooke, R. (1665). Micrographia, or Some Physiological Descriptions of Minute Bodies Made by Magnifying Glasses with Observations and Inquiries Thereupon. London: Printed by J. Martyn and J. Allestry.

Johannsen, W. (1909). Elemente der exakten Erblichkeitslehre. Gustav Fischer, Jena.

Koltzoff, N.K. (1928). Physikalisch-chemische grundlage der morphologie. *Biol. Zbl.* 48: 345–369

Laemmli, U.K. (1970). Cleavage of structural proteins during the assembly of the head of bacteriophage T4. Nature, 227(5259): 680-685.

Lane, N. (2015). The unseen world: reflections on Leeuwenhoek (1677) 'Concerning little animals.' Philosophical Transactions 370(1666): 0962-8436. https://doi.org/10.1098/rstb.2014.0344

McClung, C. E. (1901). Notes on the accessory chromosome. *Anatomischer Anzeiger* 20: 220–226.

Mendel, J. G. (1866). Versuche über Pflanzenhybriden, "Verhandlungen des naturforschenden Vereines in Brünn, Bd. IV für das Jahr, 1865, Abhandlungen: 3–47. For the English translation, see: *Druery, C.T. &*

Bateson, W. (1901)." Experiments in planthybridization (PDF). *Journal of the Royal Horticultural Society. 26: 1–32.*

Morgan, T.H. (1925). Evolution and genetics. Princeton University Press, USA.

Muller, H. J., and Prokofyeva, A. A. (1935). The individual gene in relation to the chromomere and the chromosome. Proc. Natl. Acad. Sci. US. 21: 16-26.

O'Connor, C. (2008). Discovery of DNA as the hereditary material using *Streptococcus pneumoniae. Nature Education* 1(1):104.

Passarge, E. (1979). Emil Heitz and the concept of heterochromatin: longitudinal chromosome differentiation was recognized fifty years ago. *Am J Hum Genet.* 31(2):106-115.

Roux, W. (1883). Ueber Die Bedeutung Der Kerntheilungsfiguren: Eine Hypothetische Eroerterung [German]. (Repr. ed.2018). Forgotten Books. P-24.

Schleiden, M. J. (1839). "Beiträge zur Phytogenesis." Archiv für Anatomie, Physiologie und wissenschaftliche Medicin. 1838: 137–176.

Schwann, T. (1839). Mikroskopische Untersuchungen über die Uebereinstimmung in der Struktur und dem Wachsthum der Thiere und Pflanzen. Sander, Berlin
http://staatsbibliothek-berlin.de/

Sharma, A. and Sharma, A. (1980). Chromosome Techniques: Theory and Practice.3rd edition). Butterworth-Heinemann. P724.

Sutton, W. S. (1902). On the morphology of the chromosome group in *Brachystola magna. Biological Bulletin* 4: 24–39.

Watson, J. D. and Crick, F.H. C. (1953). Genetical Implications of the Structure of Deoxyribonucleic Acid, *Nature* 171(4361): 964.

Chapter 1

Molecular Footprints of Gene in the Cradle of Nucleic Acid (DNA and RNA)

A discrete chromosomal region which is responsible for a specific cellular product and consists of a linear collection of potentially mutable units each of which can exist in several alternative forms and between which crossing over can occur.

---- James D. Watson

The identity of the chromosome as the ultimate cytogenetical entity of life at the cellular level is indisputable in the world of biological science. It maintains its structural and functional integrity which transfer from generation to generation within biological organisms. The structural variation in terms of shape, size and anatomy is evident in distinct type of cells, structurally and functionally (e.g. somatic cells and germ cells) and in divisional stages of cell division (e.g. prophase, metaphase, anaphase, and telophase of mitosis and meiosis divisions).

As transfer of hereditary characteristics are embedded in the chromosomes, the structural and ultrastructural formation of chromosomes were discovered by different researchers. Their individual observations, ideas, opinions and hypotheses in hereditary and evolutionary science and frontier research emerged from the basal node of cytogenetics in the 19th century and spread like the apical node of molecular biology in the 21st century. In the beginning of the 18th century, chromosome was considered as a cell organoid, found in the nucleus of organism's cell. There were some living organisms having large numbers of chromosomes which are very small, and are recognized as "microchromosomes" (Burt, 2002; Axelsson et al, 2005). In some of the advance groups of eukaryotic organism like avifauna and reptiles, each somatic cell contains around 30–40 pairs of microchromosomes along with 5–7 pairs of regular or normal-sized macrochromosomes. The distinction of smaller size in comparison to the regular-sized chromosomes helped to recognize microchromosomes dispersed in macrochromosomes (Fillion, 1998). The contemporary chromosomal study at the molecular level revealed that the number of microchromosomes were constant in a particular species.

In 1928, N.K. Koltzoff, who devised the molecular model of chromosome, conducted a researched on "the template principle of chromosome reproduction" and helped to understand the idea of continuity of living systems and its evolutionary journey by means of transferring genetical traits from producer to successor. In 1958, Francis Crick's "Central dogma of molecular biology"[4] became the ultimate version of the template principle of Koltzoff (1928) (template of the first order). Later, the discovery of "protein inheritance underlay the notion of steric or conformational templates (second order) for reproducing conformation in a number of proteins" (Inge-Vechtomov, 2013). The term chromosome was used after it was found condensed, chemically stained and visible (distinct during mitotic cell divisions) under the microscope. Otherwise, chromosome appear to be fibrous structures, called chromonema, which was for the first time observed in the pollen mother cell of *Tradescantia* by Baranetzsky in 1880 (Samejima et al, 2018). The term "chromonema" was coined by Vejdovsky (1912). The chromonema consists of bead-like accumulation on the chromatin thread and the distinct regions of chromosome in the meiotic phase are called chromomeres which was first time observed by Balbiani (1876) and described by Pfitzner (1882). In the state of nuclear division, each chromosome splits leads to duplication. Each half is called chromatid. It is the centromere which is responsible for separation as well as splitting of the chromosomal pair during the cell divisional cycle. Till anaphase stage, the sister chromatids in mitosis and homologues pair of chromosome in meiosis, adheres to the centromere (which form primary constriction of chromosomes). This is followed by its splitting and translocation of each half to opposite poles where each half, attached to the micro tubular spindle, is gradually pulled to opposite poles and form two daughter cells. During this microtubular-spindle-driven translocation of each half of the chromosomes, as well as chromatids, are attached to the microtubules and are called kinetochores (Suja et al, 1991). Besides primary constriction,

---

[4] Central dogma of molecular biology: Defines the flow of genetic information in the biological system. According to this principle, DNA makes RNA and RNA makes protein.

which formed the centromere location of the chromosome during the cell division, a secondary constriction was also observed in some chromosomes in the metaphase stage of cell division. This secondary constriction form a satellite-like structure chromosome, known as satellite or SAT chromosomes. It was reported for the first time by Sergei Gavrilovich Navashin, the famous Russian biologist in 1912 (Rieger et al, 1968; Gimenez-Abian et al, 2005). The secondary constriction of SAT chromosome acted like NOR or nuclear organizer region of nucleolus and is considered to take part in RNA transcription (that is copying of RNA done from DNA template, catalyzed by the enzymatic action of RNA polymerase) (Rieger et al, 1968; Gosden et al, 1981). Generally, the chromosome numbers found in the body cells or somatic cells of any organism, were diploid (2n) (like diploid chromosome profile of human being is 2n=46) which regenerated or multiplied by means of mitotic cell division. Whereas, the concerned biological organism produced reproductive or germ cells (gametes/egg) where the chromosomal number was just half of its somatic cell. This was called haploid (the haploid chromosome of human being was found to be n=23 or 22+ x/y) which was produced as a result of reduction division or meiotic division of cells. When the male and female gametes (haploids) were fused to form diploid zygote, the constancy of the unique chromosomal profile (2n) of any particular organism (belonging to a certain species) was maintained through the generations. Hence, irrespective of plants or animals, in any living organism, each chromosome in every cell appeared as a set of two, one of which came from the maternal side and one from the paternal side. These set of chromosomes were recognized as homologues chromosomes and all homologues chromosomes are eventually diploid. So, it was generally observed that chromosomal number of human species (2n=46) and the chromosomal number of apes (Gorillas, Chimpanzees, Bonobos etc.) (2n=48), remained constant generation after generation unless an interspecific hybrid evolved (Wall, 2013). The set of chromosome, found in the male/female gametes of eukaryotic, dioecious animals or plants, which determine the sex were called allosomes while the rest were recognized as autosomes (Griffiths et al, 2000; Johnson and Lachance, 2012). The variation of chromosome number in different species and the assemblage of chromosomes help to determine, regulate and transfer hereditary characteristics at a cellular and species level of the living organisms. The entire set of chromosomes found in the nucleus of the cell in any eukaryotic organism was called karyotype (Stebbins, 1950). The study of karyotypes or karyological analysis became one of the most important chromosomal study, related to cell biology and cytogenetics, which helped to determine the systematic relationship (called karyosystematics) of genetically related species and resolving the evolutionary relationships of the same (Schweizer and Ehrendorfer, 1976; Nagpure et al, 2016). Carl Wilhelm von Nageli, the Swiss botanist discovered this technique in 1842. It was in 1881, Eduward Gerard Balbiani, a famous French embryologist, who opened the chapter on molecular basis of chromosomal structures and functions. He reported that the giant-sized Polytene chromosome in the salivary glands of dipteran flies had a thousand DNA strand (Zhimulev and Koryakov, 2009). The cycle of DNA replication without cell division formed a structure with multiple chromatids that fused to form polytene chromosome. The chromomere and the inter-chromomere region contained DNA molecule of 250Å diameter in thickness (Sharma, 1985). During the interphase, the polytene chromosome appeared as parallel bands of thick and thin lines which helped the geneticists and molecular biologists to trace chromosomal mutations and taxonomic identification of different species (Alanen, 2008). Almost at the same time, in 1882, Walther Flemming, the eminent German cytologist discovered another giant chromosome (~1mm long) known as "lamp brush chromosome" extracted from the nucleus of immature eggs of amphibians, birds and insects (Morgan, 2002; Bogolyubov, 2018). Lamp brush chromosome was found to be an ideal chromosome to study fundamental structure under a simple microscope. It helped the geneticists study the function of genomes and gene expression, mapping of chromosome and DNA sequence in chromosomes which is easily observed during the mitotic stages (Morgan, 2002; John, 2003;Gaginskaya et al, 2009). However, the most popular interpretations on the structural and ultrastructural formation of chromosome are as follows (Sharma, 1985):

   a. According to the chromomere theory of chromosomal conformation, a number of granular, bead-like chromomeres are linearly attached to each other by an achromatic string to form the chromosomal thread. A group of scientists further construed that these beady structures were nothing but tiny spools of chromosomal thread. Unwinding these spools could unveil the structural entity of chromosomes in terms of constancy of structural formation, number, position etc. that prevailed through the generations and went through the process of genetic recombination.

b. On the other hand, the chromonema theory described the chromosome as a continuous, long, linear structure which went through spinalization with little stretches during each stages of cell division to come up with distinct shapes and size of chromosomal variation. A combination of chromomere and chromonema theory of chromosome was promulgated by a group of geneticists in the 19th century who considered that the beads of chromomeres were part of chromonema thread as the bio-chemical analysis (with advance chromosomal staining, along with better ultramicroscopic technologies) clearly distinguished this location from the interchromomeric regions. Furthermore, the cytogeneticists proposed that the condensed, bead-like structures of chromomeres should be considered the functional regions of chromonema as these proactive regions engaged in synthesizing nucleoprotein during the divisional cycles of cell division.

c. The alveolar theory defined chromosome as a single thread in different divisional phases like prophase, anaphase and telophase which split into double-threaded structure during metaphase stage. The cytogeneticists further suggested that during interphase (prior initiation to cell divisional cycles), chromosomes establish "lateral anastomosis"[5] to form alveoles, which in turn, form a reticulum (chromatin reticulum). This is followed by division, condensation and further divisions to form chromosomes. There are three hypotheses promulgated to explain the thread formation of chromosomes (Sharma, 1985; O'Connor, 2008):

a. Single thread theory: According to this theory, each chromosome splits in two parts during interphase and each part translocates to the opposite end during anaphase divisional state. Chromosome appears as single thread structure except in its divisional cycle of prophase and metaphase.

---

[5] Anastomosis: A connection between two cavities or passages like branching of leaf veins, blood vessels or riverine systems etc.

b.   Double thread theory: According to this hypothesis, chromosome completes one divisional cycle prior to the metaphase state. This penultimate state of metaphase is recognized as "prometaphase." The prometaphase state reaches metaphase through a 4 partite structure. The translocation of half of the 4 partite chromosome to each pole gives an appearance of 2 partite structure in anaphase state.

c.   Quadruple thread theory: According to the hypothesis of the third group of geneticists, the chromosomal structure considered as 4 partite structure in anaphase, undergoes division in the prometaphase state becoming a 8 partite state.

Later, the renowned American molecular biologist Joseph Herbert Taylor, along with two research colleagues (1957), observed the chromosomal structures in different divisional cycles. Their observation revealed that in the anaphase state, the chromosome appeared under the ultra-microscopic image as a single filament with the double helix DNA. However, a biochemical analysis with enzyme digestion techniques along with ultra-microscopic observation by the molecular biologists of the latter half of 19th and 20th century revealed the multifilamentous structure of chromosomes (Nicolini and Ts'o, 1985). The multi-fibrillar structural pattern of chromosome was observed in the metaphase stage of the divisional cycle where the thickness of the filament was 125 Å and a pair entwined (like two spiral coils integrated in each other together) to look like one unit. It maintained the same thickness to form a thread of chromosome (Nicolini and Ts'o, 1985). The experimental investigations ascertained that bipartite chromosome (two chromatids) was an individual entity. In the preliminary stages of chromosomal study in the 19th century, it was assumed that each half chromatids was a compact form of a long DNA molecule but this concept of binemy did not support the genetic recombination theory (Sharma, 1985). Rather, the uninemic concept of chromosomal theory was found to be more tenable which proposed that a single DNA helix produced the chromosome longitudinally in a compact form. The proponent of uninemic concept of chromosome further found that on the basis of reiteration of some of the nitrogenous base sequences, eukaryotic chromosomes could be categorized into three categories (Britten and Kohne, 1968; Laird, 1980):

a.   Highly repetitive DNA: It comprises base sequence repeated million times or more and constitutes 2-10% of the total DNA.

b.   Moderately repetitive DNA: It comprises base sequences repeated hundred to thousand times and it constitutes 20-40% of the total DNA.

c.   Sperm DNA: It contains a single copy of base sequence.

The presence of three types of DNA mentioned above, confirmed the uninemic concept of chromosomes; whereas, the concept of polynemy was indirectly supported by the variation of DNA content observed among the genetically close-related species having identical shape, size and number of chromosomes (Callan, 1973). The biochemical and molecular study of chromosomes revealed that the key constituent was nuclear proteins and each chromosome contained only one DNA molecule (irrespective of the molecular weight of the concerned DNA) which might be partly or entirely be genetic material (genome) of any biological organism. Along with nucleic acid, DNA was found to be structurally integrated with histone and non-histone proteins.

In most cases, lengthy eukaryotic chromosomes contain a long DNA molecule. Binding, folding and condensing of the molecule is managed by a number of proteins known as packaging proteins[6] and chaperone proteins[7] (Sharma, 1985; Demeunynck et al, 2002; Hammond, 2017). The histone protein helps DNA to maintain its structural integration, the non-histone protein maintains the assemblage of DNA in chromosomes and DNA's functional potential of transcription to synthesize RNA from DNA template and translate it to protein. The contemporary progress in the biochemical and molecular study of chromosomes concluded that ultrastructurally DNA is a matrix in which genes are arranged in a linear sequence and functionally is able to regulate the inheritance of living organisms. The molecular biologists noticed that recording and passing down of hereditary information in

---

[6] Packaging proteins: Histones, the alkaline proteins, are main component of chromatins, and its main function is regulating genes. Histones are present in the nuclei of eukaryotic cell, packed the DNA molecules, called nucleosomes. It act as spool, surrounding which the DNA molecule surrounded with.

[7] Chaperone proteins: These proteins are mainly dealt with covalent folding and unfolding of macromolecules. It primarily helped to assemblage of nucleosomes (comprised of folded histone proteins and DNA molecules)

chromosomes was a function of the DNA molecule. Moreover, the contemporary progress in molecular footprints of inheritance revealed that 99% of the DNA content at the cellular level was found in its chromosome which is responsible for controlling genetical inheritance of the biological organism. The remaining DNA was recognized as cytoplasmic DNA (either found in mitochondria in eukaryotes or plasmids in prokaryotes) which regulate cytoplasmic inheritance of the organism (Henninger, 2012).

A long way since Mendelian era, the world of science recognized that the nucleic acid, particularly the DNA molecule, is the ultimate genetic material of most living organisms by controlling the pattern of inheritance. Chromosomes play a key role in the process of reproduction and transcription (during interphase stage of cell cycle) in the intermediate stage to form mRNA (messenger RNA) from DNA and formation of protein from mRNA to protein through translation (Lehninger, 1987). Geneticists observed that light-colored euchromatin region of the chromosome was responsible for RNA synthesis during the interphase stage when the nucleofibril was in the unwinding state which made it difficult to monitor structural changes of chromosome during interphase (Stryer, 1988). Eventually, heterochromatin region of the chromosome during interphase appeared dark and was found to be abstained in RNA synthesizing process. The euchromatin region of the chromosome during interphase contained thread-like structures made up of elementary DNP (Deoxyribonucleic protein), granular structures (200-500 Å in diameter) which, in turn, was made up of RNP (Ribonucleic protein), chromatin granules and an amalgamation of the threads and granules of DNP and RNP to synthesize RNA (Hunter, 2000; Lukong et al, 2008). Further into the process, RNA became integrated with protein at the end of the interphase stage to form mRNA followed by its transfer to the cytoplasm of the concerned cell to initiate the next state of action which is called protein translation.

However, in any living system, there are two distinct types of nucleic acids found: DNA and RNA. DNA was found to be the key genetic material regulating hereditary traits of most organisms, while RNA was found to be the genetic material of some viruses which also acted as an adapter and messenger in the biological system. The earnest endeavors of Walter Jones, in 1920 to decrypt the molecular identity of nucleic acid has been initiated with his *"trinucleotide hypothesis"* when he has been engaged with the experiment, titled: *"the chemical constituent of Adenine nucleotide and of yeast nucleic acid*" (Jones, 1920). Though, the first initiative was taken in 1893 by Kossel and Neumann who discovered the presence of two purine and pyrimidine bases in the nucleic acid (Kossel and Neumann, 1893), the same observation was later supported by Steudel in 1906 when it was revealed that "each occurred in equi-molecular proportions in thymus nucleic acid" (Steudel, 1906). It was further confirmed by a Russian-American biochemist, Phoebis Levene, after he discovered ribose sugar in 1909 and deoxyribose sugar in 1929, and defined nucleic acid as a repeating tetramer in his "tetranucleotide hypothesis" in 1931 (Levene and Bass, 1931; Tipson, 1957). Between 1947 and 1951, meticulous study by Erwin Chargaff, a Hungarian-American biochemist, disclosed two important observations—deciphering the nitrogenous base-pairing of poly nucleotides, specifically in the DNA molecule, where the number of Adenine is the same as Thymine, Guanine and the varying content of Adenine (A), Guanine(G), Thymine(T) and Cytosine(C) from species to species—thereby establishing DNA as the genetic material at a molecular level (Vischer and Chargaff, 1947; Chargaff, 1951). Since the first half of the 19th century, sincere efforts of a number of molecular biologists and biochemists like Creeth et al (1947); Fischer and Helferich (1914); McCombie, Saunders and Stacey (1945); Gulland et al. (1947a; 1947b); Hotchkiss (1948); Cohn and Volkin (1951); Brown and Todd (1955); Khorana (1968) and others helped to uncover the biochemical identity of nucleotides, polynucleotides, nucleosides and their structural identities of nucleic acids. Genetic material was finally envisioned through the Watson-Crick's "double helix model of DNA" with a convincible interpretation of its structural and functional module "DNA and the central dogma" (Watson and Crick, 1953a, 1953b; Kornberg et al, 1961; Crick, 1966).

According to Levene and Jacobs (1909), after segregating protein from nucleoprotein, the nucleic acid was studied further for its building block known as nucleotide. Levene introduced the nucleotide as a unit made up of phosphate- pentose sugar (deoxyribose in DNA and ribose in RNA) and nitrogen base. Generally, there are two types of nitrogenous bases—Purines containing Adenine and Guanine and Pyrimidines containing Cytosine, Thymine and Uracil. Cytosine was present in both DNA and RNA molecules but Thymine was found only in DNA, and Uracil in RNA (Lehninger, 1987). The diagrammatic sketches of pentose sugars (Deoxyribose and Ribose sugars), and nitrogenous bases purines; pyrimidines are presented in Fig. 1.1, Fig. 1.2, and Fig. 1.3. The combination of a nitrogenous base with a pentose sugar linked by a N-glycosidic bond to form a nucleoside was recognized as:

a.   Deoxyadenosine or Adenosine

b.   Deoxyguanidine or Guanidine

c.   Deoxycytidine or Cytidine

d.   Deoxythymine or Uridine

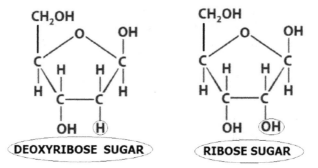

Fig. 1.1. The diagrammatic sketches of pentose sugars.

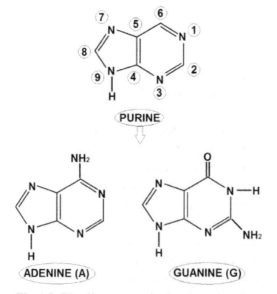

Fig. 1.2. The diagrammatic sketches of purines.

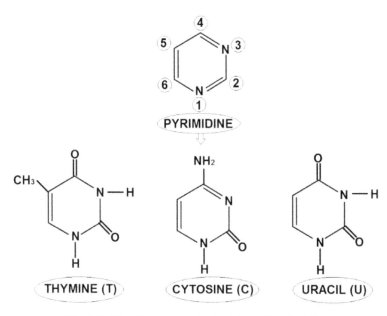

Fig. 1.3. The diagrammatic sketches of pyrimidines.

The linkage of phosphate group to the hydroxyl group (-OH) of 5' carbon position of the nucleoside by a phosphodiester bond led to the formation of the reciprocal nucleotide or dexynucleotide on the basis of the nature of the pentose sugar it integrated with (Stryer, 1988). The diagrammatic sketch of the three segments of nucleotide is presented in the fig. 1.4. When two or four nucleotides are sequentially attached to one another by the 3' carbon position of one to the 5' carbon position of the another by a phosphodiester bond, it led to form di- or tetra-nucleotide (Lehninger, 1987; Stryer, 1988; Krebs et al, 2012). Likewise, a long chain of polynucleotide was synthesized similarly. So the synthesis of polynucleotide rendered the completion of biochemical synthesis of biopolymer at one end with the free phosphate moiety of 5'position of ribose sugar. It was further recognized as 5'-end of the polynucleotide chain. Likewise, at the other end of the polymer, the ribose sugar had hydroxyl bond (-OH) at 3' position was recognized as 3' of the polynucleotide chain (Lehninger, 1987; Stryer, 1988; Krebs et al, 2012). The nitrogenous base of the polynucleotide chain was integrated by sugar moiety of the nucleotide backbone frame which formed as a result of the amalgamation of pentose sugar with phosphate molecules of the sequentially attached string of nucleotides (Lehninger, 1987; Stryer, 1988; Krebs et al, 2012). It was ascertained by Watson-Crick's double helix model that DNA molecule constituted of two inter-twined polynucleotide chains, whereas, the RNA molecule constituted of one polynucleotide chain. The diagrammatic sketch of the double-stranded polynucleotide chain is presented in fig.1.5. Molecular biologists noticed that in RNA, every nucleotide possessed an extra hydroxyl (-OH) group in the 2' position of the pentose sugar while Uracil base was found instead of Thymine (of the complementary position of DNA) (Lehninger, 1987; Stryer, 1988; Krebs et al, 2012).

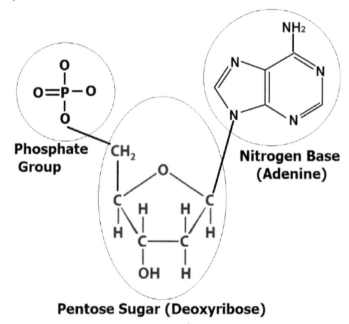

Pentose Sugar (Deoxyribose)

Fig. 1.4. The diagrammatic sketche of the three segments of nucleotide.

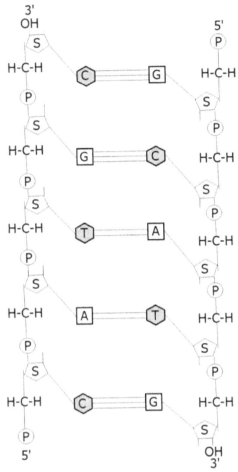

(Left) Fig. 1.5. Two-stranded polynucleotide chain.

In 1951, eminent British biochemist, Rosalind Franklin was working on the structural configuration of DNA and she successfully developed a three-dimensional image of DNA which was earlier conceptualized by renowned Kiwi-British physicist, Maurice Wilkins in Kings College, London. Franklin's crystallographic presentation intrigued two geneticists, James Watson and Francis Crick, to propound the three-dimensional model of double helix of DNA in 1953. Watson and Crick along with Wilkins was awarded the Noble prize for physiology in 1962. Unfortunately, Franklin was not alive at that time and there was no option to award posthumous research achievements till 1974. There was also a guideline for Noble prize felicitation that stated the prize money must not be shared by more than three individuals.

However, the preliminary efforts by Fredrich Meischer (1871) opened up avenues for the study of nucleic acid by identifing "nuclein" which led to the double helix model of DNA propounded by Francis and Crick (1953b) after almost a century. In the three-dimensional model of DNA, the main component was the nitrogenous base pairing of DNA between two nucleotide chains and the ratios of Adenine:Thymine and Guanine:Cytosine which was found to be constant. It was conceptualized by Chargaff (1951) that for each nucleotide of Adenine, one Thymine (A=T) base required. Likewise for each nucleotide of Guanine, one Cytosine (G≡C) base was required to be paired by hydrogen bonds. The elucidation of the base-pairing pattern of the polynucleotide double helix chain model of DNA helped to understand that the bases on the reciprocal strands were complementary to each other. If we were able to read out the base sequence of any strand (which is regarded as parent DNA strand), we would be able to synthesize the reciprocal polynucleotide chain, recognized as daughter DNA strand, made with complementary base pairs, These conclusion on genetical as well as molecular identity of DNA was universally recognized by the world of science. As per the three-dimensional model of DNA molecule, proposed by Francis and Crick in 1953, the ladder like structure of the molecule curved in a spiral with consistent gaps in between the 'steps' to form the double helix.

The diagrammatic sketch of a double helix model of DNA is presented in fig. 1.6. The key features of the spiral staircase like double-helical DNA molecule are as follows (Lehninger, 1987; Stryer, 1988; Krebs et al, 2012):

1. The two spiral ladder of the DNA molecule maintains opposite and parallel polarity. So, if a strand has to maintain polarity from 5' → 3', the reciprocal strand maintains its polarity from 3' → 5' direction. Similarly, the spiral has to maintain its right-handed rotational movement.

2. The strands of the polynucleotide chains constitute of sugar-phosphate linkage attached with the "steps" (made up of nitrogenous bases).

3. A portion of the step, in between the complementary strand, constitute of either purine or pyrimidine base. So, each 'step' is made up of a complementary base pairs—(A=T) or (G≡C).

4. The bases in two reciprocal strands bond with each other by utilizing hydrogen bonds in between the complementary base-pairs (which is purine-pyrimidine combination) specifically through two hydrogen bonds (A=T) and three hydrogen bonds (G≡C).

5. The constitutive length of the 'step,' made up of purine/pyrimidine combination, help to maintain uniform and equal distance between the strands of DNA molecule and which is found to be around 20Å in diameter.

6. The pitch of the helix or completion of one rotation of the each polynucleotide chain take a vertical, linear space of 34Å. As each of the polynucleotide chain is made up of 10 nucleotides and 10 base-pairs at every turn, the universal distance between each base-pairs is found to be 3.4Å.

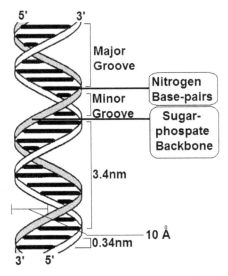

(Left) Fig. 1.6. DNA Double helix model.

The erstwhile observation in the molecular biological aspect of the base-pairing orientation in DNA molecules helped molecular biologists to establish the quantitative relationship in which (A=T) and (G≡C) or A+T = C+G or (A+T) / (C+G) totalled to 1. The tentative results indicated another hypothetical probability where the ratio of bases like A/T and G/C should maintain a ratio of 1:1 to satisfy base-pairing rules. However, contemporary studies in the field revealed that the ratio of (A+T) / (G+C) varied from species to species as (A+T) content in more complex groups of eukaryotic plants and animals had to be high in respect to (G+C) content. This observation was of immense importance which helped Karyotaxonomists and molecular phylogeneticists. Though, Watson and Crick's (1953a;1953b) simplistic explanation of the DNA in the format of "Double helix model" introduced the world of science to form an idea about the structural configuration of hereditary material, efforts of Francis Crick (1958) brought forth the functional module of the DNA: "Central Dogma of Molecular Biology" which was another milestone in the history of research in molecular biology. It flawlessly defined the function of DNA as well as affirmed the flow of genetic information from DNA to RNA during protein formation. The diagrammatic sketch of it has been presented in fig. 1.7.

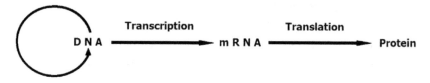

Fig. 1.7. The Central Dogma.

As can be seen, the eukaryotic DNA is a long chain of polynucleotides which gives an impression that DNA molecule of birds or mammals is very long but it does not reveal a lot about the quantitative perception of the length of the DNA molecule of higher organisms (Alberts et al, 2002). Let's find the approximate length of the DNA. It has already been established that the linear gap between two base pair of a nucleotide as well as of the polynucleotide chain is 3.4Å or .34 nm or 0.34 x 10^{-9} m. The number of base-pairs in the genome of the following organisms are:

- The number of base-pairs of the bacteriophage (Ø x 174) (The virus of the host bacterial species *E.coli*): 5,386

- The number of the base-pairs of *Agrobacterium tumifaciens*: 4,674, 062

- The number of the base-pairs of human or *Homo sapiens*: 3.3×10^{9}

- The number of the base –pairs of *Psilotum nudum*: 2.5×10^{11}

The above count of base-pairs of the entire DNA strand of the consecutive species helps us to calculate the approximate length of the unpacked DNA strand of each species, by multiplying the number of total base pairs of entire polynucleotide chains with the linear gap of each nucleotide, which is 0.34 x 10^{-9} m mentioned herewith:

- The length of the unpacked DNA molecule of bacteriophage (Ø x 174) is around 0.0000018312 m. or 1,831.24 nm.

- The length of the unwound DNA molecule of the *Agrobacterium tumifaciens* is around 0.0015891811 m or 1,589,181.09 nm.

- The length of unfurled DNA molecule of human is around 1.12 m.

- The length of unfolded DNA molecule of *Psilotum nudum* is around 85 m.

When the diameter of a nucleus of any living cellular organism is about 6μ or 10^{-6} m in average, it is inevitably puzzling that a small nucleus accommodates a DNA molecule of 1.12 m or even 85 m long. Let's go through the folding/packaging pattern of DNA helix (Alberts et al, 2002).

In prokaryotes like bacteria or viral cells, there is no well-defined nucleus, rather, its nucleus-like genetic material is known as nucleoid or genophore. It is also recognized as naked gene as it has no nuclear membrane (Day, 1991; Nowotny and Testa, 2012). The genome of prokaryotic organisms contain multiple copies of scattered round-shaped, double-stranded structures called plasmids. DNA in the nucleoids, being negatively charged, integrates with the positively charged proteins and organizes itself in loop-like formations (Alberts et al, 2002).

In eukaryotic cells, the folding/packing of DNA in nucleus is more intricate. Molecular biologists observed that protein biomolecules in the nucleus acquire positive charges through its building blocks, i.e. amino acids, specifically from the residues of the amino acid side-chain (Alberts et al, 2002). The positively charged proteins in nucleus of the eukaryotes are called histones and are comprised of amino acid residues of lysines and arginines (Annunziato, 2008). The five histone fractions integrate together to form a nucleosome. Out of the five fractions, four are H2A, H2B, H3 and H4 which come together to form histone octamer (H1 performes as a linker; H3 and H4 plays an active role in the evolution of higher group of organisms) (Alberts et al, 2002; Annunziato, 2008). It was observed that positively charged histone octamer was tightly coiled and adpressed by the negatively charged DNA helix (around 200 base-pairs) to form nucleosome (Oudet et al, 1978). Moreover, it was reported that nucleosome was attached to thread-like structure in the nucleus known as chromatin reticulum. This was visible in stained condition under microscopic view. The nucleosomes occurring in the middle of chromatin threads at frequent intervals looked like "bead on strings." This structure further coiled and condensed in the metaphase state of the cell divisional cycle (Oudet et al, 1978). Molecular geneticists observed that for the precise folding of chromatin fibers, a unique protein played a role by means of micro-space utilization, along with histone protein, which is known as non-histone chromosomal proteins or NHC (Stein et al, 1974). According to the quantitative estimation of protein and DNA content in the double helix model, the ratio of DNA to histone protein is always 1, whereas, the ratio of non-histone protein to DNA varied from

0.2 to 0.3, due to the cytogenetic activity in the tissue of a particular type of organization (Palter et al, 1979; Reeck, 1985). After viewing the stained chromatin threads under the ultra-microscope, cytogeneticists distinguished two distinct features of chromatin threads. The lightly stained, skinny and loosely packed chromatin thread, recognized as euchromatin, functionaed as proactive nuclear organoid, taking part in transcription (forming mRNA from DNA strand). The darker, densely-packed, thicker part of the chromatin thread, recognized as heterochromatin, was found in the inactive part of chromatin thread as well as chromosome.

The X-ray crystallography further unveiled that there were some exceptions to the universal double helix model of DNA, like *Fd, ØR, ØX* 174 etc. which possessed a single-stranded DNA. Biochemically, the molecular geneticists observed that in a single-stranded DNA, A and G did not form complementary base-pair with T and C bases and the non-equivalence of A and T, and G and C (e.g. the quantitative analysis of bases in the single-stranded DNA of *ØX* 174 phage revealed that the nucleotide contained 24.5% A, 32.5% T, 24.1 % G, and 18.5% C in it) were confirmed with further X-ray crystallographic analysis (Ilag et al, 1995).

Though A, T, G, C were recognized as the most common bases forming base-pairs in the formation of nucleotide as well as DNA molecule, a number of rare nitrogenous bases such as 6-methyladenine, 6-methylaminopurine, 5-methylcytosine, 5-methylhydroxycytosine etc.) were also observed later which substituted the common bases at the time of base-pairing (Stryer, 1988). The presence of unusual bases like 5-methylcytosine (MC) and 5-hydroxymethylcytosine (HMC) were found in the higher organisms (crop plants, insects, birds, mammals etc.) as either of them substituted cytosine at the time of base-pairing during DNA formation (Wyatt, 1951; Loeb and Cohen, 1959). In the DNA molecule of some bacteriophage viruses (e.g. *Ø PBS1* and *Ø PBS2*), the usual pyrimidine base (T) was replaced by Uracil which is normally found in RNA molecules (Stahl et al, 1972).

RNA is a polymer which constitute repeated units of nucleotides or polynucleotide chains in the same way the DNA polymer are made up. Though RNA has some resemblance to DNA, it is structurally and functionally different from DNA molecule (Alberts et al, 2002; Darnell, 2011):

a.   RNA is either found freely in the cytoplasm or attached to the cell organelles but DNA is mostly found in the chromatids of the nucleus organoid.

b.   The biosynthesis of RNA (transcription process) depends on the structural formation of the DNA template, whereas, DNA is a self-replicating entity.

c.   RNA constitute of Ribose sugar (a pentose sugar which looks similar to deoxyribose sugar but lacks oxygen in 2'carbon position), while DNA is made up of deoxyribose sugar.

d.   RNA consists of nitrogenous bases, Adenine (A), Guanine (G), Cytosine(C) and Uracil (U). Pyrimidine base, Uracil in RNA is replaced by the pyrimidine base, Thymine (T) of DNA. Though both are structurally almost same, the 5' carbon position of Uracil is not methylated.

e.   RNA molecule is single-stranded, in comparison to the double-stranded structure of DNA.

Though these stereotypic distinctions between RAN and DNA were repeatedly tested, single-stranded DNA was also noticed among a number of bacteriophages (e.g. *Fd*, *ØR*, *ØX* 174, *S 13* etc.). Likewise, careful examination of genetical materials of a number of organisms such as the Herpes virus, polynoma virus, T2 bacteriophage, Turnip Yellow Mosaic Virus etc. found double-stranded RNA, along with the conventional single-stranded RNA molecule which is found among most of the animals and plants (Darnell, 2011). Contemporary study of genetic materials also found DNA molecules in different organelles like cytoplasm, mitochondria, chloroplast, endoplasmic reticulum etc. along with nucleus. Pyrimidine base, Uracil, the well-recognized base of RNA was also noticed among the DNA constitution of some bacteriophage viruses (e.g. *Ø PBS1* and *Ø PBS2*). Hence, the demarcation between DNA and RNA is not so distinct. Rather, these findings raised the bar in the arena of molecular genetics to explore the structural and functional aspects of genetic material as well as the molecular basis of heredity or gene in biotechnological-driven research at the "Nano" or "Pico" level (Wang and Wang, 2014).

The following are the are three categories of RNA which carry forward
the transcription process of "Molecular basis of Central Dogma" for
biosynthesis of protein (Lehninger, 1987; Alberts et al, 2002; Darnell, 2011):

a. Ribosomal RNA or r-RNA: The r-RNA constitutes 85-90% of the
    protein in the whole organism and it is structurally made up of 70S
    ribosomal unit. This further consists of two sub-units—50S and 30S
    type, attached by magnesium ions (Shine and Dalgarno, 1974).
    Further investigation at a molecular level ascertain that the part of r-
    RNA attached to ribosome was found to be double-stranded,
    whereas, the part joined to the protein was single-stranded (Cotter,
    McPhie and Gratzer, 1967). Its specific function is yet to be
    established, though it is suggested that it assists messenger RNA to
    expedite protein translation.

b. Messanger RNA or m-RNA: The m-RNA constitute 5-10% of the
    entire RNA content in a cell. This short-lived RNA is synthesized in
    the nucleus in the presence of RNA polymerase enzyme while the
    process of transcription takes place simultaneously. The process of
    transcription uses a single, unwound strand of DNA (as a result of
    split double-strand DNA) as template and forms a reciprocal RNA
    or m-RNA strand by using complementary nitrogen bases (A,U, G
    and C) on the backbone (made up of ribose sugar and phosphate).
    During m-RNA formation, transitional structure like DNA-RNA
    hybrid strand is formed. Since it carries the message to synthesis
    protein (by means of translation process of m-RNA to protein) from
    DNA molecule it is called messenger RNA (Shine and Dalgarno,
    1977; Lehninger, 1987).

c. Transfer RNA or t-RNA: The t-RNA is the smallest RNA. It is a
    single-stranded structure with a unique folding pattern, recognized
    as clover leaf model, which has a double-stranded appearance. The
    diagrammatic sketch of t-RNA is presented in fig. 1.8. It is formed
    by "pseudotype of hydrogen bonding" between closely assigned
    bases. Towards the head or frontal end of t-RNA, there is an
    anticodon triplet, called anticodon loop which is complementary to
    the codon (triplet code or genetic code) of m-RNA responsible for
    encoding the specific amino acids. The tail-end of t-RNA is
    engaged in binding the specific amino acid encoded by the triplet

codon of m-RNA. In "Central Dogma," Francis Crick, hypothesized "Genetic code" (the triplet base sequence of m-RNA which are able to pick specific amino acid during protein chain formation) where that there must be an adapter molecule in one end to read out the code and bind specific amino acid to the other end. The adapter or anticodon loop or the amino-acid acceptor at the end of t-RNA help to bind and sequence the amino acids to form polypeptide chains and in turn, form polypeptide chains. While a number of polypeptide chains synthesize proteins at the end of translation process. For each amino acid, t-RNA is specific. Though there is a t-RNA known as initiator t-RNA but there is no t-RNA found so far that stops the process.

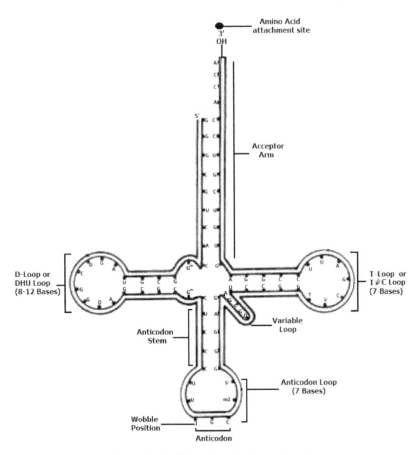

Fig. 1.8. t R N A - the adapter molecule.

In the backdrop of spatio-temporal matrix, the search for the identity of the "Mendelian factors" has gone through a number of paradigm shift. From the era of cytology followed by cytological genetics (Cytogenetics), molecular biology, biotechnology to molecular genomics, the world of science is familiar with the true identity of "nuclein" as well as nucleic acid. Contemporary research on the ultrastructural and functional aspects of DNA and RNA proved that DNA is the predominant genetic material but RNA also has properties similar to DNA (RNA is found in a number of organisms like bacteriophage *Fd*, *ØR*, *ØX* 174; Polyoma virus, Turnip yellow mosaic virus etc. as genetic material). The next concern was, if both types of nucleic acid—DNA and RNA—have the potential to be recognized as genetic material, which one is better? In order to draft a comparative account of DNA and RNA regarding their structural and functional potential as genetic material, following parameters need to be considered (Alberts et al, 2002; Calladine et al, 2004; Yarus, 2010):

  a. Possession of "Mendelian factors" and its transmission through the generations: Where coordination of protein synthesis is concerned, RNA was found directly responsible for completion of translation process in a single step. Whereas, DNA was found to be dependent upon RNA to complete protein synthesis in 2 steps. The first step involved completion of transcription (formation of m-RNA from DNA template) and the second involved completion of translation (synthesis of protein from m-RNA strand).

  b. Replication potential: As far as the replication of both type of nucleic acids are concerned, DNA replication is straight forward where 1 parent strand of the double-stranded DNA helix acts as a template after splitting and the new strand formed by sequential arrangement of the bases is complementary to the base sequence of the parental DNA strand and DNA polymerase enzyme expedite the process. Whereas, RNA replication is a DNA-depended process where RNA polymerase helps the sequential arrangement of RNA bases, complementary to the bases of the parental DNA template.

  c. Biochemical and structural stability: Observation of the "transforming principle" by Fredrick Griffith, in 1928, proved that even after the death of the host organism, the genetic material, that is the DNA molecule, remained viable. The heat-killed organism still contained viable DNA content which has ability to transform structural and functional characteristics of a sensitive organism to its resistant form. In the RNA World hypothesis, Alexander Rich (1962), has advocated that RNA, played a key role in the biosynthesis of protein on Earth which initiated the journey of life, yet, DNA has been accepted as the functionally and structurally better ingredient for the biological evolution on Earth. Structurally, the presence of (-OH) or hydroxyl bond in 2nd carbon position in RNA makes it unstable and swiftly degrading (thermos-labile) molecule due to heat application. Hence, DNA molecule has better structural stability in comparison to RNA molecule.

d. Low rate of mutation: In different situations, both DNA and RNA are subjected to mutation over course of time. From an evolutionary perspective, evolutionary biologists proposed that nucleic acid, subjected to change instantaneously in case of fast-paced mutagenesis, would be suitable to evolve faster. As RNA molecule is structurally more unstable than DNA, the biological organism containing RNA was supposed to evolve much faster than the organisms constituting the DNA molecule.

Hence, a comparison of structural and functional aspects of DNA and RNA, indicates that DNA is more structurally stable than RNA but RNA is functionally more proactive, in terms of transforming genetic information, than DNA as it directly takes part in the biosynthesis of protein. DNA, on the other hand, plays a passive role as it depends on RNA for biosynthesis of protein. It was further construed that the organisms containing DNA don't evolve as fast as the organisms containing RNA which plays a main role to steer biological evolution on Earth.

References

Alberts, B., Johnson, A., Lewis, J., Raff, M., Roberts, K., and Walter, P. (2002). Molecular Biology of the Cell. 4th edition. New York: Garland Science; Available from:
https://www.ncbi.nlm.nih.gov/books/NBK21054/

Annunziato, A. (2008). DNA packaging: Nucleosomes and Chromatin. Nature Education 1(1):26.

Axelsson, E., Webster, M. T., Smith, N. G., Burt, D. W., and Ellegren, H. (2005). Comparison of the chicken and turkey genomes reveals a higher rate of nucleotide divergence on microchromosomes than macrochromosomes. *Genome research* 15(1): 120–125.
https://doi.Org /10.1101/gr.3021305

Balbiani,E.G.(1876).Sur les phenomenes de la division du noyau cellulaire. C.R.Acad.Sci.(Paris). 83:831-834.

Britten, R.J. and Kohne, D. E. (1968). Repeated sequences in DNA. Science 161(3841): 529-540.

Brown, D. M. and Todd, A. R. (1955). Nucleic acids. Annual review of Biochem.24: 311-338.

Burt, R. S. (2002). "The Social Capital of Structural Holes." Pp. 148–92 in The New Economic Sociology, edited by Mauro F. Guille´n, Randall Collins, Paula England, and Marshall Meyer. New York: Russell Sage Foundation.

Burt, R. S. (2002). "The Social Capital of Structural Holes." Pp. 148–92 in The New Economic Sociology, edited by Mauro F. Guille´n, Randall Collins, Paula England, and Marshall Meyer. New York: Russell Sage Foundation.

Calladine, C., Drew, H., Luisi, B., and Travers, A. (2004). Understanding DNA (3rd Edition). Academic press, p.352.

Callan, H.G. (1973). DNA Replication in the Chromosomes of Eukaryotes. In: Hamkalo B.A.,Papaconstantinou J. (eds) Molecular Cytogenetics. Springer, Boston, M A. doi.org/10.1007/978-1-4615-7479-8_3

Chargaff, E. (1951). Some recent studies on the composition and structure of nucleic acids. J. Cell. Comp. Physiol. 38(1): 41–59.

Cohn, W.E., and Volkin, E. (1951). Nucleoside-5'-phosphates from ribonucleic acid. Nature. 167: 483–484.

Cotter, R., Mcphie, P. and Gratzer, W. (1967). Internal Organization of the Ribosome. *Nature* 216: 864–868. doi.org/10.1038/216864a0

Creeth, J.M., Gulland, J.M. and Jordan, D.O. (1947). Deoxypentose nucleic acids, Part III. Viscosity and streaming birefringence of solutions of the sodium salt of the deoxypentose nucleic acid of calf thymus. J. Chem. Soc. 214: 1141–1145.

Crick, F.H.C. (1958). On protein synthesis. *Symp. Soc. Exp. Biol.* 12: 138–1163.

Crick, F.H.C. (1966). Con-Anticodon pairing: The Wobble hypothesis. J Mol Bio. 19:184–191.

Darnell, J. (2011). RNA: Life's Indispinsible molecule. Cold Spring Harbor laboratory Press. P.393.

Day,S.(9ᵗʰNovember,1991).The first gene on Earth. https://www.newscientist.com/article /mg13217944-500-the-first-gene-on-earth/

Demeunynck, M., Bailly, C., and Wilson, W.D. (2002). Small molecule DNA and RNA binders: From synthesis to nucleic acid complexes. Wiley-VHC Verlag GmbH & Co., KGaA.

Fillon, V. (1998).The chicken as a model to study microchromosomes in birds: a review. *Genet Sel Evol* 30: 209. https://doi.org/10.1186/1297-9686-30-3- 209.

Fischer, E. and Helferich, B. (1914). Synthetische Glucoside der Purine. EurJIC 47(1): 210-235. https://doi.org/10.1002/cber.19140470133

Giménez-Abián, J.F., Díaz-Martínez, L.A., Wirth, K.G., Andrews, C.A., Giménez-Martín, G., and Clarke, D.J. (2005). Regulated Separation of Sister Centromeres depends on the Spindle Assembly Checkpoint but not on the Anaphase Promoting Complex/Cyclosome. Cell Cycle 4(11): 1561-1575.

Gosden, J.R., Spowart, G. and Lawrie, S.S. Satellite DNA and cytological staining patterns in heterochromatic inversions of human chromosome 9. *Hum Genet* 58: 276–278 (1981). https://doi.org/10.1007/BF00294922

Gosden, J.R., Spowart, G. and Lawrie, S.S. Satellite DNA and cytological staining patterns in heterochromatic inversions of human chromosome 9. *HumGenet* 58: 276–278 (1981). https://doi.org/10.1007/BF00294922

Griffith, F. *(1928).* "The Significance of Pneumococcal Types". *Journal of Hygiene. Cambridge University Press. 27 (2): 113–159.* doi: 10.1017/ S0022172400031879

Gulland, J.M., Jordan, D.O. and Taylor, H.F.W. (1947b). Deoxypentose nucleic acids, Part II. Electrometric titration of the acidic and basic groups of the deoxypentose nucleic acid of calf thymus. J. Chem. Soc. 213: 1131–1141.

Gulland, J.M., Jordan, D.O. and Threlfall, C. (1947a). Deoxypentose nucleic acids, Part I. Preparation of the tetrasodium salt of the deoxypentose nucleic acid of calf thymus. J. Chem. Soc., 212: 1129–1130.

Hammond T. M., (2017).Sixteen years of meiotic silencing by unpaired DNA. Adv. Genet. DOI: 10.1016/bs.adgen.2016.11.001.

Henninger, J. (1ˢᵗ October, 2012). The 99 percent….of human genome. http://sitn.hms.harvard.edu/flash/2012/issue127a/

Hotchkiss RD (1948). The quantitative separation of purines, pyrimidines, and nucleosides by paper chromatography. *J Biol Chem* 175: 315–332.

Hunter, G. K. (2000). Vital Forces: The discovery of the molecular basis of life. Academic press. 364p.

Ilag, L.L., Olson, N.H., Dokland, T., Music, C.L., Cheng, R.H., Bowen, Z., McKenna, R., Rossmann, M.G., Baker, T.S., and Incardona, N. L. (1995). DNA packaging intermediates of bacteriophage φX174. Structure. 3(4):353-363. doi: 10.1016/S0969-2126(01)00167-8

Inge-Vechtomov, S.G. (2013). The template principle: Paradigm of modern genetics. *Russ J Genet* 49: 4–9.
      doi.org/10.1134/ S1022795413010055

Johnson, N.A., and Lachance, J. (2012). The genetics of sex chromosomes: evolution and implications for hybrid incompatibility. Ann N Y Acad Sci.; 1256:E1-22. doi: 10.1111/j.1749-6632.2012.06748.x. PMID: 23025408; PMCID: PMC3509754.

Jones, W. (1920). The action of boiled pancreas extract on yeast Nucleic acid. American journal of Physiology 52(1): 203-207.

Khorana, H.G. (1968). Synthesis in the study of nucleic acids. The Fourth Jubilee Lecture. Biochem J.109(5):709-25. doi: 10.1042/bj1090709c.

Koltzoff, N.K. (1928). Physikalisch-chemische grundlage der morphologie. *Biol. Zbl.* 48: 345–369.

Kornberg, S.R., Zimmerman, S.B., and Kornberg, A. (1961). Glucosylation of deoxyribonucleic acid by enzymes from bacteriophage-infected E. coli. J Biol Chem. 236:1487–1493.

Kossel, A., and Neumann, A. (1893) Ueber das Thymin, ein Spaltungsprodukt der Nucleinsäure. Ber. Deut. Chem. Ges. 26, 2753–2756 (Quote translated in Portugal and Cohen, 1977, p. 60)

Krebs, M., Held, K., Binder, A., Hashimoto, K., Den Herder, G., Parniske, M., Kudla, J., and Schumacher, K. (2012). FRET-based genetically encoded sensors allow high-resolution live cell imaging of Ca^{2+} dynamics. Plant J. 69: 181–192.

Laird, C.D. (1980). Structural paradox of polytene chromosomes. Cell 22(3): 869- 874.

Lehninger, A. L. (1987). Principles of Biochemistry. CBS Publishers and Distriutors, New Delhi.

Levene, P. A., and Bass, L. W. (1931) Nucleic Acids, Chemical Catalog Company, New York: Available online at:
      https://babel.hathitrust. org/cgi/pt?id=uc1. b4165245;view=1up;seq=5
      (Accessed December 14, 2018)

Levene, P. A.and Jacobs, W. A. (1909). "Über die Pentose in den Nucleinsäuren" [About the pentose in the nucleic acids]. *Berichte der deutschen chemischen Gesellschaft* (in German).42(3):3247–3251. doi:10.1002/cber.19090420351

Loeb, M. R., and Cohen, S.S. (1959). The origin of purine and pyrimidine deoxyribose in *Escherichia coli* systems. J. Biol. Chem. 234:360-363.

Lukong, K.E., Chang, K.W., Khandjian, E.W. and Richard, S. (2008). RNA-binding proteins in human genetic disease. Trends Genet. 24: 416-425.

McCombie, H., Saunders, B.C. and Stacey, G.J. (1945). Esters containing phosphorus. I. J. chem Soc. (London) 1945. 380.

Miescher, F. (1871). "Ueber die chemische Zusammensetzung der Eiterzellen." Medicinisch-chemische Untersuchungen. 4: 441–460.

Nagpure, N.S., Pathak, A.K., Pati, R., Rashid, I., Sharma, J., Singh, S.P., Singh, M., Sarkar, U. B., Kushwaha, B., Kumar, R. and Murali, S. (2016). Fish Karyome version 2.1: a chromosome database of fishes and other aquatic organisms, *Database*, Volume 2016, baw012, https://doi.org/10.1093 /database/baw012

Nicolini, C. and Ts'o, P.O.P. (1985). Structure and function of the genetic apparatus. Springer link, NATO ASI Series (NSSA, Volume 98).

Nowotny, H., and Testa, G. (2012). *Geni a nudo. Ripensare l'uomo nel XXI secolo*. Torino: Codice edizioni.

O'Connor, C. (2008). Discovery of DNA as the hereditary material using *Streptococcus pneumoniae. Nature Education* 1(1):104.

Oudet, P., Germond, J.E., Bellard,M., Spadafora, C., and Chambon, P. (1978). Nucleosome structure. Philos Trans R Soc Lond B Biol Sci. 283(997):241-58. doi: 10.1098/rstb.1978.0021.

Palter, K.B., Foe, V.E. and Alberts, B. M. (1979). Evidence for the formation of nucleosome-like histone complexes on single-stranded DNA. Cell 18(2): 451-467. doi.org/10.1016/0092-8674(79)90064-3

Pfitzner,W. (1882).Ueber den feineren Bau der bei der Zeilteilung auftretenden fadenformigen Differenzierungen des Zeilkerns. Morph.Jahrb.7:289-311.

Pfitzner,W. (1882).Ueber den feineren Bau der bei der Zeilteilung auftretenden fadenformigen Differenzierungen des Zeilkerns. Morph.Jahrb.7:289-311.

Reeck, G. R. (1985). Nucleosome Structure, in Chromosomal Proteins and Gene Expression (Reeck, G.R., Goodwin, G.H., and Puigdomenech, P., (ed.) Plenum Press, New York, pp. 1-17.

Rich, A. (1962). On the problems of evolution and biochemical information transfer. In: Kasha M., Pullman B., editors. Horizons in Biochemistry. Academic Press; New York, NY, USA. pp. 103–126.

Rieger, R., Michaelis, A. and Green, M.M. (1968). *A glossary of genetics and cytogenetics: Classical and molecular*. New York: Springer-Verlag. P-533

Samejima, K., Golobdorodko, A., Gibcus, J., Dekker, J., Mirny, L and Earnshaw, W. (2018). A Pathway for Mitotic Chromosome Formation. The 43rd FEBS Congress Prague. 7-12 July, 2018. S.21-1. http://www.febscongress.org/

Samejima, K., Golobdorodko, A., Gibcus, J., Dekker, J., Mirny, L and Earnshaw, W. (2018). A Pathway for Mitotic Chromosome Formation. The 43rd FEBS Congress Prague. 7-12 July, 2018. S.21-1. http://www.febscongress.org/

Schweizer, D., and Ehrendorfer, F. (1976). Giemsa banded karyotypes, systematics, and evolution in *Anacyclus* (*Asteraceae-Anthemideae*). *Pl Syst Evol* 126: 107–148.

https://doi.org/10.1007/BF00981668

Sharma, A. K. (1985). Chromosome structure. Perspective Report Ser 14, Golden Jubilee Publications. Indian National Science Academy 1-22.

Shine, J., and Dalgarno, L. (1974). The 3'-terminal sequence of *Escherichia coli* 16S ribosomal RNA: complementarity to nonsense triplets and ribosome binding sites. *Proc Natl Acad Sci U S A.* 71(4):1342–1346.

Stahl, F. W., McMilin, K.D., Stahl, M.M., and Nozu, Y. (1972). An enhancing role for DNA synthesis in formation of bacteriophage lambda recombinants. Proceedings of the National Academy of Sciences 69 (1972): 3598–601. http://www.pnas.org/content/69/12/3598.full.pdf (Accessed December 9, 2016).

Stebbins, G. L. (1950). *Variation and evolution in plants.* Oxford University Press, London, UK.

Stein, G. S., Spelsberg, T. C., and Kleinsmith, L. J. (1974). Nonhistone chromosomal proteins and gene regulation. Science. 183 (4127): 817-24.doi: 10.1126/science.183.4127.817.

Steudel, H. (1906). Die zusammensetzung der nucleinsäuren aus thymus und aus heringsmilch. Hoppe-Seyler Z. Physiol. Chem. 49: 406–409 10.1515/bchm2.1906.49.4-6.406

Stryer, L. (1988). Biochemistry. W.H. Freeman & Co. Ltd. 1089p.

Suja, J.A., de la Torre, J., Giménez-Abián, J. F., García de la Vega. C., and Rufas, J. S. (1991). Meiotic chromosome structure. Kinetochores and chromatid cores in standard and B chromosomes of *Arcyptera fusca* (Orthoptera) revealed by silver staining. Genome. 34(1): 19-27. doi: 10.1139/g91-004. PMID: 1709128.

Suja, J.A., de la Torre, J., Giménez-Abián, J.F., García de la Vega. C., and Rufas, J.S. (1991). Meiotic chromosome structure. Kinetochores and chromatid cores in standard and B chromosomes of *Arcyptera fusca* (Orthoptera) revealed by silver staining. Genome. 34(1):19-27. doi: 10.1139/g91-004. PMID: 1709128.

Taylor, J.H., Woods, P.S., and Hughes, W.L. (1957). The organization and duplication of chromosomes as revealed by autoradiographic studies using tritium-labeled thymidinee. Proc Natl Acad Sci U S A. 43(1):122-8. doi: 10.1073/pnas.43.1.122.

Tipson R. S. (1957) Phoebus Aaron Theodor Levene, 1869–1940: Obituary. Adv. Carbohydr.Chem. 12: 1–12.

Vejdovsky, F., (1912). - Zum Problem der VererbunF}trager. Kgl. Bohm Ges. Wiss. Prag.

Vejdovsky, F., (1912). - Zum Problem der VererbunF}trager. Kgl. Bohm Ges. Wiss. Prag.

Vischer, E. and Chargaff, E. (1947). The separation and characterization of purines in minute amounts of nucleic acid hydrolysates. J. Biol. Chem. 168:781-782.

Wall. J. D. (2013). Great ape genomics. *ILAR journal*, 54(2): 82–90.

https://doi.org/10.1093/ilar/ilt048

Wang, E.C., and Wang, A. Z. (2014). Nanoparticles and their applications in cell and molecular biology. Integr Biol (Camb). 6(1):9-26. doi: 10.1039/c3ib40165k.

Watson, J.D. and Crick, F.H.C. (1953a). A structure for deoxyribose nucleic acid. Nature. 171:737–738.

Watson, J. D. and Crick, F.H. C. (1953b). Genetical Implications of the Structure of Deoxyribonucleic Acid, *Nature* 171(4361): 964.

Wyatt, G.R. (1951). Recognition and estimation of 5-methylcytosine in nucleic acids. Biochem. J. 48 (1951): 581-584.

Yarus, M. (2010). Life from an RNA world. Harvard University Press. P 194.

Chapter 2

Exploring the Molecular Basis of Inheritance Driven by Nucleic Acid (DNA and RNA)

There are two RNA worlds. The first is the primordial RNA world, a hypothetical era when RNA served as both information and function, both genotype and phenotype. The second RNA world is that of today's biological systems, where RNA plays active roles in catalyzing biochemical reactions, in translating mRNA into proteins, in regulating gene expression.

---- Thomas R. Cech

The role of nucleic acid and the dominance of DNA in proving the universal identity of molecular inheritance of life has remained unequivocally indisputable in the history of biological evolution. Though the double-stranded DNA was regarded as the universal genetic material when the rational outlook was to define molecular basis of inheritance, the contemporary, introspective visualization of biological evolution and its turn over with the passage of time pushed us to look at the role of "inconsequent nucleic acid," that is, the RNA. The contemporary "RNA World hypothesis" was a composite outcome of the brilliant compilation of ideas, convincing and fascinating introspections of hypotheses and interpretations of experimental observation regarding the structural and functional aspects of nucleic acid, particularly, RNA's which helped to view the biological history of evolution of life on Earth from a new perspective. Though the term "RNA World" has formally coined by Walter Gilbert, an American biochemist and Nobel laureate in 1986, the "RNA World hypothesis" was independently proposed earlier by Carl Woese, Francis Crick and Leslie Orgel in 1960s, immediately after the discovery of the double-helical structure of DNA (Woese, 1967; Crick, 1968; Orgel, 1968; Cech, 2012). In 1962, the great American biologist/biophysicist, Dr Alexander Rich of Harvard Medical School strongly advocated for "RNA World hypothesis" for the first time, on the basis of RNA's direct involvement in protein synthesis (Neveu et al, 2013). The DNA molecule might be structurally more stable than RNA molecule, but it's indirect protein synthesis or protein translation mechanism involved an extra-step called "transcription" which rendered formation of RNA from DNA and formation of protein which depended on the translation potential of the RNA molecule. Till the beginning of 80's, there was a misconception regarding the bio-catalytic properties of the biochemical and organic molecule as the proteins (enzyme molecules) were considered to be biocatalysts. In 1980, independent research on the bio-catalytic properties of organic molecules, conducted by Sidney Altman and Thomas Cech, found that RNA molecule acted as a unique biocatalysts which expedite the biochemical reactions. Going one step forward, in 1989, Altman and Cech, propounded the catalytic RNA molecule as "Ribozymes" and the first ribozyme was recognized as "Tetrahymena ribozyme" (Zimmer, 2014). The discovery of ribozyme was regarded as a milestone in the history of nucleic acid study, particularly the RNA which was introduced as the storehouse of genetic information and catalyst of biochemical reaction which drove the biological evolution in the "Primordial RNA World" as well as begin the journey of life on Earth. The existence of a unique organelle— "Ribosome"— in the cell and the giant molecular complex constituting RNA and protein was recognized as ribozyme. However, the structural and functional review of the Ribozyme molecule showed that it was not the structural constituent of ribosome but RNA was responsible for the protein translation process. Such a discovery of biochemical potential of RNA strengthened the "RNA World Hypothesis" by the evolutionary biologists to construe the primitive eukaryotic and early eukaryotic organisms which evolved with RNA as inherent genetic material. Genetic material among a number of virus, primitive microscopic organisms (those that structurally look like non-living organisms and inorganic/organic out of the biologically active or living in the host bodies) and functional living/biological organisms have pointed towards the RNA as the early genetic material or primitive molecular basis of inheritance which supposedly shifted the paradigm from inorganic/organic evolution to biological evolution on the Earth. Nowadays, it is a proven fact that a number of important biochemical, physiological and cyto-genetical functions like metabolism, protein translation, gene splicing etc. have been integrated to some extent with RNA molecule which acts like genetic material back up for the journey of life on

Earth. "RNA World hypothesis" is strongly supported in the contemporary works of a number of evolutionary biologists, such as:

a.   The origin of RNA world: Co-evolution of genes and metabolism (Copley et al, 2007).

b.   "The strong RNA world hypothesis: Fifty years of old" (Neveu et al, 2013).

c.   "Common origins of RNA, protein and lipid precursors in a cyanosulfidic proto-metabolism" (Patel et al, 2015). etc.

According to the "RNA world hypothesis," RNA molecule evolved to be the most dominating, self-replicating nucleic acid on Earth, before the evolution of DNA and proteins. The molecular biologists observed that RNA could contain and replicate genetic information with the help of ribozymes, as it was done by protein enzymes in DNA molecules. As RNA molecules were found to be fragile and short-lived, it may have not played a positive role in the process of biological evolution. Rather, it could have taken a long duration of time to complete the process. Hence, molecular biologists suggested that "some ancient RNAs may have evolved the ability to methylate other RNAs to protect them" (Rana and Ankri, 2016).

It was noticed that RNA catalysts were not functionally stable like protein enzymes of DNA, rather, it was found to be biochemically reactive and structurally unstable. Evolutionary biologist suggested that the primordial evolution of the RNA world, supposedly, originated and evolved with "Ribonuleoprotein." The Ribonuleoprotein went through organic evolution and diversification of life on Earth in quest of a genetic molecule which had better storage and protein-synthesizing abilities and helped to evolve "Deoxyribonucleoprotein," giving rise to the structurally stable and functionally conservative DNA molecule in course of time. Logically, 2 is always better than 1 when there is a possibility of change in shape of any genetic entity of a living organism. It might affect the functional pattern of genetic expression of the concerned organism which could change in course of time, thereby, altering the entire course of biological evolution. From this perspective, the evolutionary advantage of possessing the double-stranded DNA molecule by an organism over the single-stranded RNA molecule as genetic material is supposedly a better choice as the organisms with DNA have better stability against potential mutational changes and correctional/

repairing opportunities in post-mutation regime. Thus, in the dynamic age of Anthropocene, RNA provides better opportunity for a biological species to evolve faster, whereas, the DNA ensures stability and diversification in an ever-changing environment.

However, as genetic material, the sustainability of DNA molecule as a superior nucleic acid was unequivocally appreciated than RNA in the world of science. Along with the proposition of double-helix model of DNA, Francis and Crick proposed a method for its replication, *i.e.* formation of new strands from parental DNA strands. Molecular biologists found that complementary base-pairing was the most unique characteristic that attributed structural stability of the nucleotides to the polynucleotide chains. If the experimental studies of linear sequential arrangement of nitrogenous bases for the nucleotide in a polynucleotide chain was found to be: CATTGACTAA, then the complementary base sequence in the other chains would be: GTAACTGATT. The consecutive double hydrogen bond between A=T and triple hydrogen bond between G≡C helped to hold the nucleotide pair together, as well as, keep the structural integrity of two parallel strands of DNA molecule as unique genetic material. According to Watson and Crick's (1953a) DNA replication model, the helically-coiled strands of DNA would merely separate and act as a pair of template (complementary to one another) for synthesizing the complementary strands. It was also observed that breaking of hydrogen bonds between the complementary bases between polynucleotide strands was followed by separation of the two strands and unwinding of each strand, leading to the formation of DNA template for duplication. So, the bases in the parental template strand were involved in complementary base-pairing (like Adenine paired with Thymine and Guanine paired with Cytosine base). Hence, the complementary base-pairing along the parental template followed by its integration with the sugar-phosphate (deoxyribose) backbone lead to the formation of new strands.

Meselson and Stahl (1958), who studied the DNA structure and its replication in *Escherichia coli*, strongly advocated in favor of Francis and Crick's DNA replication model and further observed that the newly formed strand of DNA was always found to be complementary to the old DNA template strand, which acted as a primer. Hence, this mode was recognized as semi-conservative method of DNA replication. Along with semi-conservative mode of DNA replication, there are two more patterns of DNA replication that was proposed (Griffith et al, 2000):

a. Conservative: According to this model, newly-synthesized strands form a copy of the parental strand, where the parental strand remains unchanged.

b. Dispersive: According to the model, the double helical parental strand breaks down randomly into segments. Each segment replicates itself followed by random attachment among the newly synthesized and old parental segments to form a patchwork of DNA.

A diagrammatic sketch of semi-conservative type of DNA replication is presented in fig. 2.1.

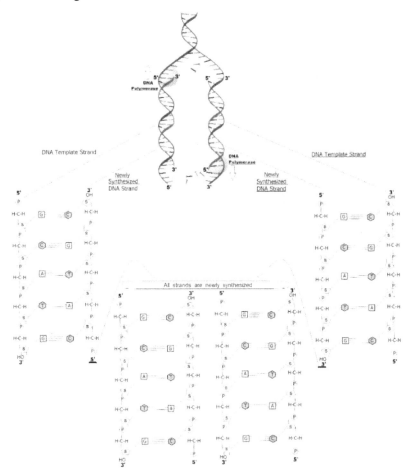

Fig. 2.1. Semiconservative replication of DNA.

Consecutive studies on DNA replication in higher group of plants and animals confirmed the semi-conservative model of DNA replication further. Like any biochemical reactions, the DNA replication process was assisted by a number of biological catalysts or enzymes. The key catalyst which took part in DNA replication process was recognized as DNA polymerase. DNA polymerase enzyme used the un-coiled DNA strand as template and involved polymerization of a big number of nucleotides as well as deoxy nucleotides in short span of time. The rate of DNA replication varies from species to species, for example, the DNA replication of the prokaryotic bacterial DNA of *Escherichia coli* which has double-stranded circular DNA or plasmid or single DNA could be referred here (Meselson and Stahl, 1958). Molecular biologists recorded that it took 40 minutes to replicate around 4.7 million base-pairs which shows that the rate of replication of DNA in prokaryotes to be around around 1000 base-pairs/second (Bank, 2017).

It was also observed that the process of replication was initiated from one end of the DNA strand to the opposite and the complementary sequencing of the nitrogenous bases was done precisely (as the chance of error was found to be limited to one in billion times). Any flawed replication of polymerization would lead to further gene mutations which would burn the bridge of inheritance of consistent genetical and structural identity transfer from ancestor to successor. It would attribute the cytogenetic consistency of any organism to replicate itself as well as survive with ancestral biological features for generations. Unless, natural selection process favored certain derived cytogenetic features, there would be a chance of emergence of genetic inconsistency at species level that would either be deleterious, leading to the evolution of a new entity of life or a dead end of existing species. In case of eukaryotic DNA replication, the process proved to be complex. Human chromosome contain 150 million base pairs and each individual has 46 chromosomes which usually replicates at the same time— during S-phase of cell division (Bank, 2017). Technically, each cell of the human species is supposed to be involved in DNA replication at the rate of 50 base-pairs / second and is supposed to take more than a month for a cell to replicate a chromosome but it has been noticed that in reality, it took only an hour or less to complete the process. This is done by the coordinated multiple replication process in different parts of the chromosome at the same time along with synchronised enzyme actions (like stitching different segments of the cloth at the same time) to get the job done faster and without flaw (Bank, 2017).

The basic biochemical principle of the enzyme on the substrate is the functioning idea of lock and key, hence it was evident that the enzyme, DNA polymerase, worked upon the substrate (nucleoside), Deoxy nucleoside triphosphates, toward its ultimate goal to carry out DNA replication. At a cellular level, any biochemical action was driven by the energy fuel "ATP" or Adenosine-Tri-Phosphate, likewise, the power-supply for DNA replication was supplied by the terminal ends of the phosphates of "Deoxy nucleoside triphosphates" (Stillman, 2013). Let's consider the strategic aspects of DNA replication where we could consider the analogy of a road-repair model. If we repair any one of a two-way lane, we could either break the entire (damaged) lane to repair it which would involve time, a huge workforce, money and the diversion of traffic through a single lane leading to traffic issues for a long period of time or we could do the repair work in small areas, then we could manage it with limited resource and divert the traffic by making a loop with less trouble. Since unwinding an entire double-stranded DNA by disintegrating all hydrogen bonds between the two polynucleotide strands would consume a long time and a huge amount of energy, the unwinding of an entire thread of double helical stranded DNA, involving its replication, through a small opening of the parental DNA helix, recognized as "replication fork" (like unzipped end) was more feasible. It needs to be mentioned that the disintegration of the bonds between the complementary bases was catalyzed by the enzyme "Helicase" (McGlynn, 2013).

The progress of DNA polymerization had been driven by DNA-dependent polymerase, and it has involved in a complementary synthesis of new DNA strand against the parental strand of the 3' to 5' polarization; simultaneously identical course of action was witnessed from other end of the DNA strand, recognized to be reciprocal polarization, observed from 5' to 3' direction, and in the language of molecular biological perspective, the bio-synthesis of DNA molecule had been contemplated to be "continuous" but "mix and match combination" course of action of polarization and reciprocal polarization (Brewer, 1975). While, on the other strand, where the polarization of the parental DNA strand ran from 5' to 3' direction, the polymerization of a complementary new strand gradually synthesized in the opposite direction in "discontinuous" or fragmented form. These discontinuous synthesized fragments of DNA was biochemically joined by the enzyme action of DNA ligase (Brewer, 1975; Hyone Myong, 1996).

Moreover, it was observed that the polymerization of DNA was not initiated randomly at any point, rather, it was a small location (also called "vector"), which initiated a specific segment of the parental DNA strand known as "Origin of replication" (del Solar et al, 1998).

After traversing the trail of molecular biology, DNA replication is still an enigmatic process of the "Central Dogma in Molecular Biology" where DNA replication in eukaryotes as well as in higher group of animals and plants, take place in the interphase or "S-stage" before the organism is ready for the cell divisional cycle (Crick, 1958). It was also found that erroneous polymerization led to mutation of DNA strand which would lead to either extinction of an extant species or evolution and diversification of a new species from the extant genetic mother-stock. Likewise, the disruption of cell division immediately after DNA replication or duplication would lead to the formation of polyploid organisms (double or triple diploid set of chromosome). Such genetic anomaly of the new/derived line of organisms would result in the evolution of a unique species or formation of a genetically non-compatible organism having little or no sexual reproductive potential.

Let's go to the next functional level of "Central Dogma in Molecular Biology" (Crick, 1958). At this level of processing genetic information, the copying of genetic information from DNA strand to RNA strand took place which had some functional resemblance to the "DNA replication" process, and is known as RNA transcription. In this process, the complementary base-pairing was sequentially arranged to synthesize RNA, as well as messenger RNA (mRNA) with respect to the parental strand of DNA. The key process of transcription was mainly witnessed in the complementary Purine-Pyrimidine base-pairing of RNA as DNA replication involved Adenine (Purine) – Thymine (Pyrimidine) pairing. While, in the synthesis of RNA molecule, specifically mRNA, pyrimidine-base Uracil has been formed instead of Thymine, complementary to the DNA parental strand. When DNA duplication involved the copying process of the entire DNA of the concerned organism, it was noticed to be a two-step process, initially begin with unwinding of the parental double helix DNA into two distinct strands , followed by the complementary replication of each strand of the entire length of the DNA of parental strand was formed, so the newly-synthesized pair of strands out of two parental DNA strand was found to be a complimentary copy of the parental DNA strand. However, RNA transcription involved synthesis of a small segment of single RNA strand complementary to the DNA molecule. The transcription involved synthesis of a RNA strand, complementary to a single DNA strand but it never synthesized both the DNA strand. Being a single-stranded molecule, it is quite inconvenient to form complementary RNA strands for two DNA strands as the sequential arrangement of the base sequence of both the RNA strands would be different in such a situation. Certain RNA bases involved in sequential arrangement of amino acids are linked to each other by peptide bonds to form polypeptide chains or specific proteins. So, it's clear that for coordination of uniform protein synthesis or translation process, synthesis of uni-stranded RNA is necessary out of one DNA strand as transcription of double-stranded RNA (which is not a usual RNA structure) might lead to the formation of a chaotic poly-peptide chains as well as erroneous forms of protein. It would also stall the third-lap of the "Central Dogma in Molecular Biology," as single-stranded RNA is involved herewith in protein translation where the potential role of two RNA strands was found to be absolutely hypothetical. Hence, the tentative/hypothetical formation of double-stranded RNA in the hypothetical transcription process, as well as the involvement of the same DNA molecule would be responsible for formation of diverse nature of proteins (at least two or more proteins) that does not comply within the uniformity of gene expression. Such hypothetical attempt of transcription might jeopardize the protein translation process.

However, molecular biologists noticed three operational locations of DNA (Gaughan, 2016):

a.   Promoter

b.   The structural gene

c.   Terminator

As the double-stranded DNA helix has two reciprocal polarities (one strand runs from 5' to 3' direction, while the other runs from 3' to 5' direction), it's difficult to distinguish the two strands which act like the potential DNA template strand to transcribe RNA molecule. In order to sort out this convoluted issue, molecular biologists searched for the particular DNA strand with distinct polarity (that could carry-out the transcription process), in relation to its response to the DNA-dependent RNA polymerase enzyme.

The molecular biologists observed that the DNA-dependent RNA polymerase enzyme carries out polymerization from 5' to 3' direction which indicated that the enzyme worked with respect to the DNA strand having 3' to 5' polarity, the template strand. The template strand, under the catalyzing effect of RNA polymerase, starts synthesizing single-strand RNA. In this regard, it should be mentioned that the reciprocal DNA strand of the template strand (from 3' to 5' direction) whose polarity is from 5' to 3' direction was recognized as the "coding strand" (Dotson, 2018). Apparently, the coding strand ran from 5' to 3 end and acted like a reference for RNA strand (with the exception of Uracil base instead of thymine base). Structurally, the coding strand of a DNA acts like mirror image of mRNA strand or RNA transcript (where Thymine is replaced by Uracil). The diagrammatic pattern of the DNA template or Anti-sense strand and DNA coding or Sense strand (Dotson, 2018) is presented in fig 2.2.

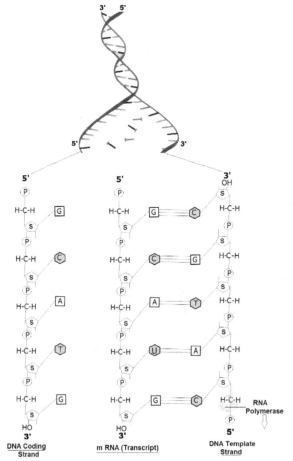

(Left) Fig. 2.2. Diagrammatic sketch of DNA template strand, DNA coding strand and RNA transcript.

The transcribed RNA, as well as the structural gene, has two distinct ends, "promoter" at 5' polar location and the "terminator" located towards the 3' flank (Gaughan et al, 2016). Hence, the promoter's location of the structural gene is recognized towards 5'-end location of coding strand, called upstream, while the terminator's location is recognized at the 3'-end location of coding strand, known as downstream. When initiation of the transcription process has been taken care off by the promoter in the upstream region, the end of transcription process has been completed with the terminator at the downstream location of the coding strand of DNA. The diagrammatic sketch of promoter, and terminator, flanking the structural gene is presented Fig. 2.3.

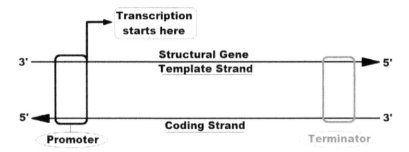

**Fig. 2.3. The diagrammatic sketch
of structural gene, flanked by Promoter and Terminator.**

Contemporary studies on molecular footprints of RNA synthesis help us to understand the gene action, however, the earliest endeavor to define RNA synthesis was started with Ochoa (1961), who observed the transitional role of RNA was transcribing information stored in DNA and translating it through the synthesis of protein. It was generally observed that the transcription process in eukaryotic organisms was more complicated than prokaryotic organisms (Clancy, 2008). The in-depth structural and functional aspects of three RNA polymerase catalysts showed that three species of RNA molecules were catalyzed by three RNA polymerase enzymes. Out of the three, one RNA polymerase enzyme was localized in the nucleolus region and was found to be responsible for synthesizing ribosomal RNA or rRNA. While the other two RNA polymerase enzymes were found to be localized outside nucleolus but within the nucleus and these two RNA polymerase catalysts were responsible for synthesizing messenger RNA (mRNA) and transfer RNA (tRNA).

Though it was unequivocally proven that gene is the ultimate structural and functional unit of inheritance and life, it was difficult to physically locate it as a structural entity of DNA. Though, protein translation process (the third lap of "Central Dogma in Molecular Biology"), involved DNA sequences, coding for RNA (particularly ribosomal and transfer RNA)

helped to define the functional identity of gene. Before digging deeper for
molecular footprints of gene in the matrix of DNA sequences a penultimate
perception of the Gene-enzyme relationship needs to be reviewed. In an
effort to discover this relationship, a number of researches were carried out
on a diverse group of micro-organisms since the beginning of the 19th
century. It started with a fungal mold "*Neurospora crassa*" by George W.
Beadle and Edward L. Tatum in 1941, who defined it as "gene-enzyme"
relationship. However, it was Norman H. Horwitz, an eminent biochemist,
as well as geneticist who propounded the "One gene-one enzyme"
hypothesis. A brief outline of the "One gene-one enzyme hypothesis" is
mentioned herewith.

Beadle and Tatum (1941) started experimenting with a wild strain of
Neurospora, that was able to grow in the "minimal medium" containing
sucrose, minerals, nitrate and the vitamin content of biotin and the potential
to produce most of the amino acids required for proteins. The treatment of
mutagen on the wild strain of *Neurospora* led to the production of next
generation of mutant *Neurospora* which could not grow in the minimal
medium unless supplemented with amino acids Citrulline and vitamin
niacin. Hence, it was inferred that a mutant strain of *Neurospora* failed
biosynthesis of amino acid Citrulline. Likewise, another mutant strain of
Neurospora was isolated which was unable to biosynthesize Arginine (an
amino acid). The cross between the mutant strain and the wild strain
revealed that there was a single difference between them that would be able
to block biosynthesis of an amino acid of any mutant strain produced by its
erstwhile wild strain. However, three mutant strains of *Neurospora* were
screened to find those that could successfully grow on minimal media with
supplements of arginine or Citrulline or ornithine. The detailed biochemical
analysis of the pathways of the formation of protein starting from Ornithine
revealed that there were two steps to convert Ornithine to Citrulline and
Citrulline to Arginine, while another mutant strain had only one step to
synthesis Arginine from Citrulline. The critical reviews of the biochemists
ascertained that the three consecutive steps from Ornithine to Citrulline,
from Citrulline to Arginine and from Arginine to protein required three
specific enzymes for specific blocks (Newmeyer, 1962; Flint and Kemp,
1981; Davis and Ristow, 1983). It indicated that the functional expression of
gene or inheritance distinguished between wild and mutant and each mutant
strain by the absence of a specific enzyme among the three enzymes or any

specific enzyme and modified the gene in those consecutive steps. So, the ratio between gene and enzyme with regards to expression of gene was found to be 1:1. The earlier perception of biosynthesis of enzyme produced by gene was categorically disregarded by the molecular biologists.

The erstwhile one gene-one enzyme hypothesis was modified later in 1945 after molecular biologists noticed that non-enzyme proteins were also encoded by gene along with polypeptide chain. Consequently, one gene-one enzyme was modified to "One gene-one polypeptide hypothesis" where it was accepted that biosynthesis of one amino acid was controlled by one gene. In reality, higher group of organisms or eukaryotic organisms, expression of any specific character was supposed to be constituted of a number of proteins and each protein was constituted of a series of amino acids. Keeping this in mind, if one amino acid is likely to be encoded by a gene, then a number of genes are responsible for expressing a character which does not fit the Mendelian concept of gene. Later, Seymour Benzer, the American physicist/molecular biologist, who meticulously studied as well as reviewed the identity of gene as the ultimate entity of inheritance in lower group of organisms came up with the "concept of gene-cistron, muton, and recon" in 1962. Benzer and Champe's (1962) observations revealed that a number of triple base pairs in DNA molecule were responsible for synthesizing few amino acids constituting of few polypeptide chains. They were found to be sequentially positioned to each other in a certain order and acted as a functional unit within a chromosome and was recognized as "Cistron." Hence, Benzer and Champe's (1962) identification of cistron— which was found to be responsible for polypeptide—had close identical resemblance to the Mendel's "factor of inheritance" as well as gene (Sharma, 1985). Furthermore, Benzer and Champe (1962) observed that the cistron was complex in nature as it constituted of a fairly large number of triplets, so the functional comparison of cistron with gene helped with the recognition of gene as a "chemically long molecule" and the ultimate unit of such a chemically long molecule was a "single base" (Sharma, 1985). This "single base" was found to be responsible for gene mutation and was called as "muton" by Benzer and Champe (1962). The molecular biologists noticed that the "single base" status of muton was found in all organisms, though the linear size of cistron varied (within chromosome) as it constituted a number of triplets which corresponded with the orientation of polypeptide chains (Sharma, 1985). To define it simply, cistron is that part of the DNA which is responsible for encoding a polypeptide and is considered to be the structural unit of gene. Contemporary studies on the molecular structural footprints of mRNAs among prokaryotes and eukaryotes clearly distinguished that mRNA of prokaryotes were constituted of more than one cistron, hence called poly-cistronic, whereas the mRNAs of eukaryotes or advance group of organism was found to be mono-cistronic (Blumenthal, 2004). Similarly, "Recon" was identified as the ultimate unit of genetic recombination and the linear size of it was found to be varying (Benzer and Champe, 1962). So, reviewing the identity of gene as well as the "Mendelian factor" as the ultimate unit of inheritance forced molecular biologists to step out of the stereotypical box of "omnipotent functional unit of inheritance," after recognizing the structural and functional entity of gene at its molecular level in the form of cistron, muton and recon discovered by Benzer and Champe (1962).

During the biosynthesis of protein at a cellular level, the DNA-dependent RNA polymerase acts as a common biocatalyst on 3 distinct species of RNA to expedite protein translation (Lodish et al, 2000), and which are as follows-

a. Messenger RNA (mRNA): mRNA is responsible for forming template copied out of DNA complementary strand.

b. Transfer RNA (tRNA): tRNA is involved with reading genetic code and organizing the amino acids on its way to protein translation.

c. Ribosomal RNA (rRNA): rRNA plays a distinct role in catalysis and structural organization in the final phase of protein translation.

It was observed and clearly stated by the molecular biologists that DNA-dependent RNA polymerase enzyme initiate the RNA polymerization in prokaryotes (mainly bacteria) by sequencing and synchronizing complementary base-pairs. It is complementary to the 5' polar end of the reciprocal DNA template and the sigma factor (ᕗ factor) which assists to tag RNA polymerase enzyme to the promoter region and initiate transcription or RNA polymerization (Belfort et al, 1995). RNA polymerase enzyme was found to act upon nucleoside triphosphate as the substrate to undertake RNA polymerization by complementary base sequencing that transcribed RNA strand in the opposite direction of DNA template strand. The mechanism, however, is not clear but it was revealed that RNA polymerase kept transcribing RNA thread and lengthening it by adding complementary sequence of base pairs against the DNA template once the б factor initiated the polymerization at promoter region till the end. During the phase of elongation, a segment of RNA was found to be integrated to the RNA polymerase enzyme the rho factor (ρ factor) terminated the polymerization of the RNA when the elongated length of the RNA thread reached the end. At the termination phase of transcription, the RNA molecules detached from the RNA polymerase enzyme and dissociation made the RNA polymerase enzyme non-functional.

Though there is no doubt that DNA molecule contain genes, it is difficult to render a visual interpretation of gene in the language of DNA sequence. The action of the "DNA-dependent RNA polymerase" driven by RNA polymerization was found to be much more complex in eukaryotes in comparison to prokaryotes. Molecular biologists noticed that in the advance group of organisms, structural and functional role of RNA polymerase, the key enzyme catalyst taking part in RNA polymerization, was cumbersome as there are two different types of RNA polymerase enzyme on the basis of its origin. The occurrence of regular RNA polymerase was found to be integrated to the main cellular organelle—ribosome—but the existence of three sub-species RNA polymerase enzymes were found in the nucleus of a cell had distinct functions (Carter and Drouin, 2009) and those are as follows:

a. RNA polymerase I: It is responsible for transcription of diverse groups of ribosomal RNAs or rRNAs (5.8S, 18S, and 28S rRNA s).

b. RNA polymerase II: It is responsible for transcription of the heterogeneous nuclear RNA or hnRNA, the precursor of mRNA.

c. RNA polymerase III: It transcribes a number of diverse RNAs like-transfer RNA or tRNA, small nuclear RNAs or snRNA and 5S ribosomal RNA or 5S rRNA.

According to recent experimental findings, gene has been recognized as a bunch of translated sequences of DNA called exons and untranslated sequences of DNA recognized as introns. Though, exons and introns were found in the primary transcripts they were in non-functional state. From then on, eukaryotic transcription has gone through a unique genetic maneuvering at a molecular level called "splicing" where intervening introns are removed and the exons are joined together to be functionally active. Exons were found in processed (or mature) RNA, whereas, the introns—the interspersing sequence of DNA—were not. The structural pattern of exon and intron gives the gene a more complex appearance which was recognized as the "Split-gene model." The idea of split-gene came up in 1977 when molecular biologists started studying the molecular footprints of the mammalian virus recognized as adenovirus. During their investigation, Richard J. Roberts and Phillip Sharp, two eminent molecular geneticists, observed that the DNA sequences of the adenovirus, engaged in encoding a polypeptide, was not an intact thread; rather it was split in segments. This unique discovery was awarded the Nobel prize in the field of physiology or medicine in 1993. The genes in such structural orientation was recognized as split genes, or introns or junk DNA etc. The diagrammatic sketch of split-gene process is presented in fig. 2.4.

A cistron constituted of the total number of exons present in a gene which was translated into polypeptide. It was observed that in the penultimate phase of gene transcription, hnRNA—a pre-mRNA structure—was produced by a molecular/biochemical pathway recognized as "alternative splicing." This was coordinated by "Spliceosome" which drove out all the introns and gathered the exons together to form cistron (Padgett et al, 1986; Le Hir et al, 2000). The hnRNA employed a process called "capping"[8] and "tailing" to transform itself from intermediary state to mRNA. The processed or transformed hnRNA, recognized as mRNA, dissociated from the nucleus and attached itself to the micro-unit of endoplasmic reticulum in order to initiate protein translation (Sharp, 2005). The mRNA combined with ribosome and formed polysome which left the bases free to actively participate in protein translation process.

---

[8] Capping: Methyl Guanosine Triphosphate has been added to the 5' pole of hnRNA. Tailing: Multiple adenylate residues have been added in the 3' pole of hnRNA, which has been recognized as polyadenylation.

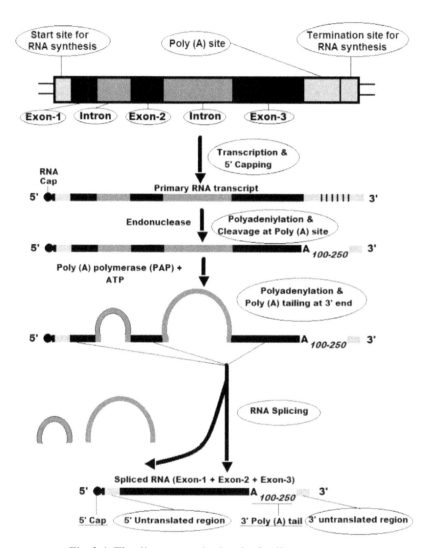

Fig. 2.4. The diagrammatic sketch of split gene process.

Hence, a gene, particularly in eukaryotes, would successfully synthesize many polypeptides (by virtue of transcription, guided by monocistronic mRNA), to provide the eukaryotic organisms with a diverse array of proteins. It was further suggested that the role of gene as a hereditary material to transfer a trait from one generation to the next one was influenced by promoter and regulatory sequence of the structural gene or cistron. Often, the regulatory sequence of the structural gene was

erroneously called the "regulatory gene," but biologists affirmed that the
"regulatory sequence" could not transcribe any RNA or encode any protein.
Hence, in recent history of molecular genetics, discovery of cistron as the
functional unit of gene ushered in a change in paradigm from "the regime of
one gene = one enzyme" or its modified outcome – " the era of one gene =
one polypeptide" to "one cistron = one polypeptide." The progress of
science has never stopped with a particular discovery, rather, the history of
scientific discoveries and innovation has always been dynamic with a spatio-
temporal trail of contemporary findings where there is no existence of
absolute truth. "True findings" of today might be an "obsolete theory"
tomorrow. So, the perception of gene as the ultimate unit of heredity, as well
as biochemical footprint of life and as a undivided or continuous thread in
DNA, was turned on its head with the discovery of split-gene. It was proved
that gene existed in segmented forms and could coordinate specific function
in such form. Apparently, the emergence of "Split-gene model" has
questioned the validity of the erstwhile "Central Dogma in Molecular
Biology" as the molecular footprint of inheritance (Sharp, 2005). For the
evolutionary biologists, the split-gene model, specifically the presence of
introns in eukaryotic DNA and its involvement in RNA polymerization, was
supposed to be an ancestral/primitive characteristic in the eukaryotic
organisms. The molecular footprints of RNA polymerization indicated tan
evolutionary bias towards the "RNA world hypothesis" along with
evolutionary origin and divergence of life where DNA might be considered
as a structurally sustainable molecular basis of inheritance. RNA would be
the functionally pro-active molecular basis of inheritance which played a
cardinal role in the journey of biological evolution and was found to be valid
till date.

References

Bank, E. [Leaf group] (2017). "List Ways in which Prokaryotic and Eukaryotic DNA Differ." List Ways in which Prokaryotic and Eukaryotic DNA Differ | Education – Seattle PI. Seattle PI, 21 Jan. 2014. [Web. 23 Apr. 2017.]

Beadle, G. W. and Tatum, E. L. (1941). Genetic control of biochemical reactions in *Neurospora. PNAS, 27*:499-506. Retrieved from http://www.pnas.org/ content/27/11/499.full.pdf.

Belfort, M., Reaban, M. E., Coetzee, T. and Dalgaard, J. Z. (1995). Prokaryotic introns and inteins: a panoply of form and function. J Bacteriol. 177(14):3897-903. doi: 10.1128/jb.177.14.3897-3903.1995.

Benzer, S., and Champe, S. P. (1962). A change from nonsense to sense in the genetic code. Proc Natl Acad Sci U S A. 48(7):1114-21. doi: 10.1073/pnas.48.7.1114.

Blumenthal, T. (2004). Operons in eukaryotes. Brief Funct Genomic Proteomic. 3(3):199-211. doi: 10.1093/bfgp/3.3.199.

Brewer, E. N. (1975). DNA replication by a possible continuous-discontinuous mechanism in homogenates of *Physarum polycephalum* containing dextran. Biochimica et Biophysica Acta (BBA) - Nucleic Acids and Protein Synthesis. 402(3): 363-371.

Carter, R., and Drouin, G. (2009). Structural differentiation of the three eukaryotic RNA polymerases. Genomics. 94: 388–396.

Cech, T. R. (2012). "The RNA worlds in context." Cold Spring Harbor Perspectives in Biology. 4 (7): a006742. doi:10.1101/cshperspect.a006742

Clancy, S. (2008). RNA transcription by RNA polymerase: prokaryotes vs eukaryotes. Nature Education 1(1):125.

Copley, S.D., Smith, E. and Morowitz, H.J. (2007). The origin of the RNA world: co-evolution of genes and metabolism. Bioorg Chem. 35(6):430-43. doi: 10.1016/j.bioorg.2007.08.001.

Crick, F.H.C. (1958). On protein synthesis. *Symp. Soc. Exp. Biol.* 12: 138–1163.

Crick, F.H. (1968). The origin of the genetic code. Journal of Molecular Biology. 38 (3): 367–79. doi:10.1016/0022-2836(68)90392-6

Davis, R. H., and Ristow, J. L. (1983). Control of the ornithine cycle in *Neurospora crassa* by the mitochondrial membrane. J Bacteriol. 154(3):1046-53. doi: 10.1128/JB.154.3.1046-1053.1983.

del Solar, G., Giraldo, R., Ruiz-Echevarría, M.J., Espinosa, M. and Díaz-Orejas, R. (1998). Replication and control of circular bacterial plasmids. Microbiol Mol Biol Rev. 62(2):434-64.

Dotson, D. (26th April, 2018). Differences between coding and template strands. https://sciencing.com/differences-between-coding-template-strands-10014226.html

Gaughan, S., Johnson, R., Wang, J., Wachholtz, M., Steffensen, K., King, T., and Lu, G. (2016). The complete mitochondrial genome of the shoal chub, *Macrhybopsis hyostoma*, Mitochondrial DNA Part B, 1:1, 911-912, DOI: 10.1080/23802359.2016.1197069

Griffiths, R.I., Whiteley, A.S., O'Donnell, A.G., and Bailey, M.J. (2000). Rapid method for coextraction of DNA and RNA from natural environments for analysis of ribosomal DNA- and rRNA-based microbial community composition. Appl Environ Microbiol. 66(12):5488-91. doi: 10.1128/aem.66.12.5488-5491.2000.

Hyone Myong, E. (1996). Enzymology Primer for Recombinant DNA Technology. Academic press. P.702.
https://doi.org/10.1016/B978-0-12-243740- 3.X5000-5

Flint, B.F. and Kemp, H.J. (1981). Cross-pathway Control of Ornithine Carbamoyltransferase Synthesis in *Neurospora crassa*. Journal of General Microbiology 128: 1503-1507.

Le Hir, H., Izaurralde, E., Maquat, L. E., and Moore, M.J. (2000). The spliceosome deposits multiple proteins 20-24 nucleotides upstream of mRNA exon-exon junctions. EMBO J. 19(24): 6860-9. doi: 10.1093/emboj/19.24.6860.

Lodish, H., Berk, A., Zipursky, S. L., Matsuaira, P., Baltimore, D. and Darnell, J. (2000). Molecular Cell Biology. 4th edition. New York: W. H. Freeman. Available from:
https://www.ncbi.nlm.nih.gov/books/NBK21475/

McGlynn, P. (2013). Helicases at the replication fork. Adv Exp Med Biol. 767:97- 121. doi: 10.1007/978-1-4614-5037-5_5.

Meselson, M. and Stahl, F.W. (1958). The replication of DNA in *Escherichia coli*. Proceedings of the National Academy of Sciences. 44 (7) 671-682; DOI: 10.1073/pnas.44.7.671

Neveu, M., Kim, H. J. and Benner, S. A. (2013). The "strong" RNA world hypothesis: fifty years old. Astrobiology. 13 (4): 391–403.

Newmeyer, D. (1962). Genes influencing the conversion of citrulline to arginosuccinate in *Neurospora crassa*. J. Gen. Microbiol. 28:215- 230.

Ochoa, S. (1961) Enzymatic Synthesis of Ribonucleic Acid. In: Aisenberg A.C. et al. (eds) Radioactive Isotopes in Physiology Diagnostics and Therapy / Künstliche Radioaktive Isotope in Physiologie Diagnostik und Therapie. Springer, Berlin, Heidelberg.
https://doi.org/10.1007/978-3- 642-49761-2_31

Orgel LE (Dec 1968). Evolution of the genetic apparatus. Journal of Molecular Biology. 38 (3): 381–93. doi:10.1016/0022-2836(68)90393-8

Padgett, R.A., Grabowski, P.J., Konarska, M.M., Seiler, S. and Sharp, P.A. (1986). Splicing of messenger RNA precursors. *Annu Rev Biochem*. 55:1119–1150.

Patel, B.H., Percivalle, C., Ritson, D.J., Duffy, C.D. and Sutherland, J.D. (2015). Common origins of RNA, protein and lipid precursors in a cyanosulfidic protometabolism. Nat Chem. 7(4):301-7. doi: 10.1038/nchem.2202.

Rana, A. K., and Ankri, S. (2016). Reviving the RNA World: An Insight into the Appearance of RNA Methyltransferases. *Frontiers in genetics*, 7:99.

https://doi.org/10.3389/fgene.2016.00099

Sharma, A. K. (1985). Chromosome structure. Perspective Report Ser 14, Golden Jubilee Publications. Indian National Science Academy 1-22.

Sharp, P. M. (2005). Gene "volatility" is most unlikely to reveal adaptation. Mol Biol Evol 22:807–809.

Stillman, B. (2013). dNTP concentration and DNA replication. Proceedings Nat. Acad. Sci. 110 (35) 14120-14121; DOI: 10.1073/pnas. 1312901110

Watson, J.D. and Crick, F.H.C. (1953a). A structure for deoxyribose nucleic acid. Nature. 171:737–738.

Woese, C.R. (1967). The genetic code: The molecular basis for genetic expression. Harper & Row, p. 186.

Zimmer, C. (25th September, 2014). "A Tiny Emissary from the Ancient Past." The New York Times. https://www.nytimes.com/2014/09/25/science/a- tiny-emissary-from-the-ancient-past.html

Chapter 3

Postulating Universality of the Genetic Code: It's Enigmatic Evolution from Ancestral Form

...the standard alphabet exhibits better coverage (i.e., greater breadth and greater evenness) than any random set for each of size, charge, and hydrophobicity, and for all combinations thereof. In other words, within the boundaries of our assumptions, the full set of 20 genetically encoded amino acids matches our hypothesized adaptive criterion relative to anything that chance could have assembled from what was available prebiotically.

---- G. K. Philip & S. J. Freeland

Two of the most important bio-molecular mechanisms in molecular genetics are replication and transcription as these two instrumental processes streamline the copying of similar or different type of nucleic which helps to posit "gene" as the universal identity of the molecular inheritance of life. As long as the concept of nitrogenous bases are clear along with its unique role of purine-pyrimidine base-pairing complementarity in the biosynthesis of nucleic acid, it's easy to understand "complementarity" as a cardinal interphase that would define replication and transcription of nucleic acids. Empirically, translation of protein involves interpretation or transposition of genetic information from the polymer of DNA nucleotides via RNA nucleotides which are translated to amino acids and are sequentially organized to form polypeptides as well as protein molecule. As there are 20 amino acids responsible for the synthesis of proteins, complimentarity between the language of 4 bases of DNA is a given. Apparently, the "complementarity" interphase is visibly non-existent between nucleotides and amino acids but it's scientifically incorrect to acknowledge visible non-existent entity as functionally non-existent entity. There are plenty of evidence in the field of biochemistry and molecular genetics, observed by scientists from time to time, that change (structural and functional) in genetic material as well as nucleic acid molecule led to change in sequencing of amino acids in the protein molecule. This laid the foundation for the concept of genetic code which stated that functional interface coordinates the passage of genetic information from RNA nucleotides to sequence the amino acids at the time of biosynthesis of protein.

In the field of applied biology, intensive study on biochemical and molecular footprint of the nucleic acids, particularly, DNA have made a tremendous progress mainly due to the involvement of biochemists, organic chemists, molecular biologists. Likewise, a wide group of scientists from inter-disciplinary fields (biochemistry, organic chemistry, biophysics, biostatistics, molecular genetics etc.) were involved in deciphering the genetic code—the most enigmatic tool in the translation process. In 1954, George Gamow, the eminent physicist as well as a scientist in quantum mechanics, strongly advocated that in order to code 20 amino acids certain permutation-combination of 4 bases (which forms nucleotides of DNA) was necessary (Gamow, 1954a; 1954b). According to mathematical calculations, a combination of 2 of 4 bases or 4^2 or 16 duplets would not be enough to code 20 amino acids (Gamow, 1954b; 1954c; 1954d). So, it was concluded that triple of 4 bases or 4^3 or 64 triplets would be enough to code 20 amino acids and such a theoretical assumption was true (Gamow, 1954a; 1954d; 1954e). These 64 triplets were called codons and it was further confirmed by the contemporary studies of Marshall W. Nirenberg (Nirenberg and Matthaei, 1961), and Hargobind Khorara (Khorana, 1979) in genetic code.

Nirenberg and his colleagues investigated *Escherichia coli* to study the genetic code in-vitro (Nirenberg and Mathaei, 1961). They used RNAase enzyme extract to synthesize the polynucleotide chain, the poly-U chain or polyuridyclic chain. In the next level, this enzyme was added to tRNAs and 20 amino acids separately. Eventually, Nirenberg and his colleagues were able to synthesis a single amino acid, i.e. Phenylalanine or polyphenylalanine. They concluded that the amino acid Phenylalanine was encoded by the polynucleotide sequence "UUU." Likewise, they found "AAA" code for Lysine, "CCC" code for proline and "GGG" code for Glycine. Though, initially, sequencing of nucleotide triplets and encoding a particular amino acid was easy, gradually it became harder when its multiple permutation-combination of triples concerned 65 triplets. In order to do that, Khorana initiated a different experimental protocol in which he used synthetic DNA or synthetic polydeoxyribonucleotides and synthesized messenger polyribonucleotides or mRNA of intended sequences for cell-free synthesis of polypeptides. In the next level, the amino acid sequence were determined by studying the sequence of the polypeptides. Khorana observed that repeated sequence of C and U (either UCU or CUC) lead to synthesis of leucine and serine, whereas, a sequential combination of A, C, and U base could form ACUACUACUACUACU. The three possible combination of triplets could form CUA, ACU, and UAC leading to the synthesis of chain of proteins (amino acids)—threonine, arginine and glutamic acid etc. Khorana (1979) was able to discover 25 codons, coding diverse amino acids like:

  a.   CUC – codes for leucine

  b.   UCU – codes for serine

  c.   ACG – codes for threonine

  d.   CGA – codes for arginine

  e.   GAC – codes for glutamic acid etc.

Based on the composite works of different molecular biologists like M. W. Nirenberg, P. Lader, O. Ochoa, H. Ghosh, I. Gupta, T. M. Jacob, R.D. Wells, P. Lengyl, J.F. Speyer etc. a table of 68 triplets or codons and its anticodons have been deciphered in terms of amino acids and is presented in the form of checker board in Table 3.1.

Exploring Genome's Junkyard:
In the labyrinth of evolution

Codon	Anti-Codon	Amino acid	Codon	Anti-Codon	Amino acid	Codon	Anti-Codon	Amino acid	Codon	Anti-Codon	Amino acid
UUU	AAA	PHE	UCU	AGA	SER	UAU	AUA	TYR	UGU	ACA	CYS
UUC	AAG	PHE	UCC	AGG	SER	UAC	AUG	TYR	UGC	ACG	CYS
UUA	AAU	LEU	UCA	AGU	SER	UAA	AUU	STOP	UGA	ACU	STOP
UUG	AAC	LEU	UCG	AGC	SER	UAG	AUC	STOP	UGG	ACC	TRP
CUU	GAA	LEU	CCU	GGA	PRO	CAU	GUA	HIS	CGU	GCA	ARG
CUC	GAG	LEU	CCC	GGG	PRO	CAC	GUG	HIS	CGC	GCG	ARG
CUA	GAU	LEU	CCA	GGU	PRO	CAA	GUU	GLN	CGA	GCU	ARG
CUG	GAC	LEU	CCG	GGC	PRO	CAG	GUC	GLN	CGG	GCC	ARG
AUU	UAA	ILE	ACU	UGA	THR	AAU	UUA	ASN	AGU	UCA	SER
AUC	UAG	ILE	ACC	UGG	THR	AAC	UUG	ASN	AGC	UCG	SER
AUA	UAU	ILE	ACA	UGU	THR	AAA	UUU	LYS	AGA	UCU	ARG
AUG	UAC	MET	ACG	UGC	THR	AAG	UUC	LYS	AGG	UCC	ARG
GUU	CAA	VAL	GCU	CGA	ALA	GAU	CUA	ASP	GGU	CCA	GLY
GUC	CAG	VAL	GCC	CGG	ALA	GAC	CUG	ASP	GGC	CCG	GLY
GUA	CAU	VAL	GCA	CGU	ALA	GAA	CUU	GLU	GGA	CCU	GLY
GUG	CAC	VAL	GCG	CGC	ALA	GAG	CUC	GLU	GGG	CCC	GLY

Table 3.1. The list of codons and anti-codons coding for various amino acids.

The key features of the genetic codes are as follows:

a. The codon is a triplet meaning each codon consists of 3 bases of mRNA strand, which codes for one amino acid. So, there are 64 codons or triplets found in the dictionary of genetic codon, of which 61 codons are engaged in coding 20 amino acids and rest do not engage in any coding, rather, function as stop codons.

b. Genetic codes are specific and unambiguous as one codon encodes one amino acid. For example, codon AUG encodes the amino acid methionine, codon UGG encodes amino acid tryptophan. A particular codon can not code for more than one amino acid.

c. A brief review of the table of genetic code shows that a number of amino acids have been encoded by more than a number of codons. For example, amino acid serine has been encoded by the codons-UCA, UCC, UCU, and UCG which helps to identify degenerate genetic code. Degeneracy is supposedly caused by a variation in the third base of the triplets recognized by the nucleotide at 5' position of anticodon and it has followed the principle of Wobble hypothesis. Hence the "Wobble hypothesis" defines that how the multiple codons encode one amino acid and it elucidates further that the tRNA (responsible for attaching amino acid), has found to be attached with more than one codons mainly due to an inherent "less precise base pairing" actions between the 3rd base-pair position of codon and the 1st base-pair of anticodon (Alberts et al, 2008).

d. According to the sequential arrangement of bases to form codons, it is clear that no base of one codon is shared by the following codon which ascertained that genetic codes are non-overlapping.

e. The codon is read uninterrupted in mRNA and there is no punctuation to be found in between two codons. The second codon starts immediately after first codon, no base point out the end or beginning of new codon.

f. By and large, genetic code is universal. For example, the codon UUU has encoded the amino acid Phenylalanine universally, from tiger to unicellular bacteria. Though, contemporary molecular biological investigations found that codons in mitochondrial are not always universal.

g. During protein synthesis experiments, it was noticed that amino acid methionine initiated the amino acid chain formation or protein synthesis, so AUG codon was considered as the initiator. Similarly, termination of polypeptide chain was regulated either by UAA (Ochre) or UAG (Amber). The term termination codon was discovered by Brenner et al (1965). Later, Har Govind Khorana, the great American molecular geneticist discovered UGA which was also considered as termination codon (Khorana, 1968).

Once the dictionary of genetic code book was ready, there be an interface that should encode the language of codons to the language of nucleoids and translate it to amino acids which is effective from nucleic acid to protein direction. Since protein, as well as amino acid, do not have any special structural configuration; mRNA codon, which has any special configuration such that any interface found to be effective from protein end screens out the reciprocal triplet sequence of codons. Hence, Francis Crick (1958) promulgated soluble RNA (sRNA) or tRNA as the potential molecular adapter which would read the sequence of codon of mRNA from one end and held the specific amino acid at the other end. Crick arranged sequential arrays of the amino acids leading to the formation of polypeptide chains and protein chains in the long run. It was found that the loop region of the tRNA adapter successfully formed an anticodon loop which was found to be complementary to the sequence of codon loop of mRNA. It was the "amino-acid-acceptor end" which bound specific amino-acid at the other end and sequentially lined them up to form polypeptide chains. It needs to be clearly understood that each tRNA is found to be specific to a particular amino acid depending upon the triplet base sequence of anticodon. Initially, a specific tRNA begins protein translation; though, no tRNAs have been recognized as stop-codons so far. The secondary structure of the tRNA was propounded as a clover-leaf model by Robert W. Holley (1966). The diagrammatic sketch of the tRNA, the adapter molecule and a visual presentation of protein translation is presented in Fig. 3.1 and Fig. 3.2 respectively.

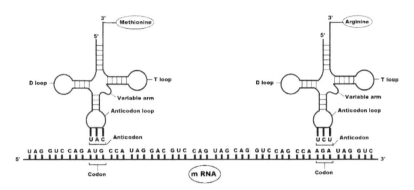

Fig. 3.1. The diagrammatic sketch of t-RNA adapter molecule.

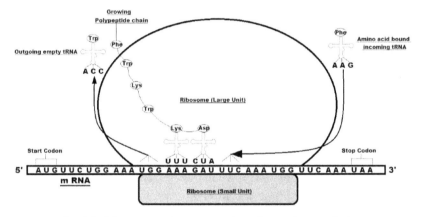

Fig. 3.2. Diagrammatic presentation of protein translation.

Molecular biologists have always referred to mutation studies to understand and define the relationship between genes and DNA. Mutation is the change or permanent alteration of the nucleotide sequence (DNA sequence) of a genome which happened either due to mistakes at the time of DNA replication or was influenced by environmental factors (physical or biochemical) e.g. UV radiations, toxic smokes, bright sunlight etc. As a result, the configuration of the DNA profile of the concerned organisms often led to genetic variations at the species level and changed the protein translation process to yield different type of proteins. It was observed that a result of deletion or rearrangement of a segment of the DNA was the mutation effect of the existing DNA profile of the concerned organisms which have gone through changes and the altered DNA profile might lose. It was also observed by immunobiologists that as a result of point mutation, a single base-pair in the gene for beta globin chain of hemoglobin has gone through changes; furthermore that has led to the structural changes of amino acid residues from glutamate to valine. Such changes in amino acid residue in the post-mutational stage cause a disease in human beings called "Sickle cell anemia." The effect of point mutation entailing deletion or insertion of bases could easily be studied from the following typographical exercise:

a.   Insertion exercise:

CAT   HAS   ONE   RED   EYE

Now if the letter 'W' is inserted between the words 'one' and 'red,' the rearrangement of the sequence of triplet would change the above statement to:

CAT   HAS   ONE   WRE   DEY   E

Instead of one, if letters 'W' and 'E' are inserted, it would read:

CAT   HAS   ONE   WER   EDE   YE

Now insertion of three letters 'W,' 'E' and 'E' would the modify statement to:

CAT   HAS   ONE   WEE   RED   EYE

Hence, after inserting three letters or a three-letter-word, the newly formed statement has a different meaning than the original sentence/statement.

In the same way, deletion exercise with the letters 'R,' 'E,' and 'D' would lead to a new arrangement with new words:

b.   Deletion exercise:

The deletion of the letter 'R' would lead to the formation of the following statement:

CAT   HAS   ONE   EDE   YE

Followed by removal of 'E':

CAT   HAS   ONE   DEY   E

Removal of 'D' would lead to:

CAT   HAS   ONE   EYE

So, it has been observed from both sets of exercise (set-A and set-B), that regardless of insertion or deletion, two bases of the triplet would change the sequence of the entire reading frame of the codons and lead to mutational changes in the nucleotides. It has also been observed that any changes in reading frames would lead to change in the meaning of the original statement, in a similar way that the mutational changes of reading frames of triplet sequence would alter the protein translation process. Hence, this kind of mutational changes which takes place due to insertion of any base is called "frameshift insertion." Whereas, any mutational changes due to deletion of any base or bases is recognized as "deletion mutation." It has generally been stated that as the key amino acids are encoded by more than one codons, there is an inherent safeguard while DNA sequencing against mutational changes. The recent progress in the field of molecular genetics has ascertained that the extra-nuclear DNA, plasmid DNA, phagemid DNA and mitochondrial DNA act like a potential molecular mechanic to correct frameshift or deletion mutations to keep the protein translation process unaltered at the species level through the generations (Tsuchiya et al, 2005).

However, the molecular geneticists also noted that protein translation can be completed in two steps:

a. Proactive role of ribosomes in charging (aninoacetylation) of tRNA: Ribosomes, the ultimate location of protein biosynthesis was found to be dissociated in two sub-units (30S and 50S) in the cytoplasm when not involved in protein synthesis. The messenger RNA attached itself to 30S sub-unit of ribosome which eventually carried AUG or start codon in the presence of the protein factor (F_3). The first tRNA carrying methionine is called f-met-tRNA f^{met} and is integrated with 30S-mRNform A complex for the formation of protein factor (F_1). During this state, 50S sub-unit ribosome is attached to 30S-mRNA complex (to form 70S unit of composite ribosome) in the presence of protein factor (F2). Each ribosome has two cavities: "P" or polypeptidyl site and "A" or Amino-acyl site. Once re-organization of ribosomal complex is complete, the elongation of polypeptide chain is coordinated in three following stages:

1. Integration of Amino acid (AA) and tRNA at the A site (amino-acyl site) of ribosome.

2. Formation of peptide bonds.

3. Translation of peptidyl tRNA from A site to P site.

During AA-tRNA integration to the A site of ribosome, the TF 1 (Transfer factor 1) and Guanosine triphosphate or GTP (originally released as GDP) was involved in the amino-acetylation of tRNA.

The next level formation of peptide bond occurred when free carboxyl group (-COOH) of peptidyl tRNA was found to be at the P site and free amino (-NH2) was found at the A site grouped with AA tRNA. Later, the charged AA-tRNA translocated from A site to P site of ribosome, driven by GTP and TF II (transfer factor 2). In the entire process, occurrence of a relative movement of ribosome and mRNA was noticed and it was due to availability of next codon in action for A site. As the ribosomes moved from one codon to the next, along mRNA molecule, the amino acids were added one by one to form polypeptides which was guided by mRNA and ultimately controlled by DNA. At the end of the translation phase of protein, a "release factor" was integrated at the stop codon to stop translation. It released the complete polypeptide chain from the ribosome. The formation of the protein chain led to the disintegration of the composite ribosomal unit of 70S structure into two sub-units 30S and 50S.

b. Formation of polypeptide chains: A specific tRNA (complementary base triplet to mRNA) anticodon pair up with mRNA reciprocal codon at one end and, carries specific amino acid to the other. Sequentially, the second tRNA poised at the next site of mRNA by means of codon-anti-codon pairing, result in sequential lining-up of amino acids connected to each other by peptide bonds. Once the first tRNA finished carrying amino acid, it becomes free in the cytoplasm to grab another amino acid to get it sequentially lined up by maintaining the amino acid logistics. So, in the translation process, pairing of mRNA and tRNA, sequencing and polymerization of amino acids leads to synthesis of protein. The term translation has been used in this process as the sequence of anti-codon is similar to the sequence of DNA molecule. When transcription occurs in the nucleus region, translation of proteins takes place in the cytoplasmic area. Molecular biologists redefined mRNA based on its mode of function as they considered that an mRNA which has potential to produce one particular protein or enzyme (and which is dependent on DNA) should also be considered as "gene" functionally. The coordination of the entire protein translation process is similar to the top-down functional pyramid, where DNA is at the top followed by the functional group such as mRNA, followed by tRNA at the base, engaged in biosynthesis of proteins/enzymes. Though mRNA is flanked by start and stop codon and involved in the formation of polypeptides, there are some extra sequence in mRNA called "unistranded regions or UTR" either present before start codon at 5' or after stop codon at 3' of RNA strand and according to the contemporary perception, this UTR has efficiently coordinated protein translation.

Though hypotheses on origin and evolution of genetic codes were developed time and again by different molecular geneticists along with the concept of nuclear acids and nitrogenous base arrangements on its way to polynucleotides formation, yet, genetic code was found to be universal or almost so. In support of it there are two main theories:

- Stereochemical theory: Woese (1967) promulgated this theory and was supported by Pelc and Welton (1966). It suggested that there were specific stereochemical fit between the amino acid and the base sequence of its codon of the concerned tRNA. It was further seen that codon sequence perceived on tRNA somewhere, but experimental observation ascertained that no such structures occurred in the tRNA for tyrosine either in *E coli* or *S. cerevisiae* (Madison et al, 1966; Goodman et al, 1968). As this theory was propounded even earlier than Crick's model work (laid out in 1967) because its difficult to ensure whether it was correct or not and if correct, up to what extent.

- Frozen Accident theory: It was promulgated by Cree in 1968. As per this theory, the genetic code was to be universal as any change in structure and function could be lethal or might be selected against. As the genetic codes defined the amino acid sequences of highly evolved protein molecules of diverse group of organisms, a small structural change needed to be counteracted by correctional mutation of the changed codes to its erstwhile forms. So continuity of any species without changed genetical configuration or phenotypic traits and its evolutionary journey on earth ascertained that the code does not change very often. It further confirmed that diverse array of life evolved from the phylogenetical mother stock called "LUCA" (Last Universal Common Ancestor) and diverged in course of spatio-temporal changes. It was further recognised that allocation of codons to distinct amino acids were not predestined but merely, a matter of chance.

According to "The Origin of Genetic Code" by Crick(1968), it's a difficult learning process to study the origin of genetic code without touching the difficult aspects of origin and evolution of protein synthesis and to reach at the root of the primitive approach of the protein synthesis process. So it is plausible that for a time being, the biochemists were "stuck by the considerable involvement of non-informational nucleic acid" (Crick, 1968). In the biosynthesis of protein, scientists found ribosome which stemmed out of RNA (tRNA and rRNA) and played a role but Crick construed (1966) that if ribosomes (as autonomous cellular organelles) were to synthesize proteins, the yield would be less and the rate of production would be slower, rather than being integrated to RNA. The origin and evolution of protein synthesis seem to be integrated to the RNA World hypothesis. In the ancient phase of biological evolution as RNA assisted protein translation would be faster, RNA itself, was very competent to organize its structural pattern in compact form. If there was no choice between DNA and RNA, instead, the only choice was shortly-persisted RNA material to any biological organism as biochemical unit of inheritance of life, then in order to achieve evolutionary sustainability, nature must have gone through a series of trial and error mechanisms and RNA driven "random protein biosynthesis" was likely to be winner of that primitive endeavour. Moreover, Crick (1966) further proposed that in the earliest efforts of RNA-driven primitive protein synthesis, no catalytic role of polymerase enzyme was present (as there was no protein). Most likely, the minerals played potential role in the early efforts of polymerization of protein synthesis. Crick (1968) further suggested that, most likely, RNA itself acted as replicase enzyme and started working as precise polymerization of amino-acid chains. This hypothesis was supported by the collaborative works of Carl Woese, Francis Crick and Leslie Orgel in 1967 (Woese, 1967).

Crick (1968) shared his perception on the structural pattern of primitive nucleic acid, which fit in the "random biosynthesis" of protein model. Along with his colleagues Woese and Orgel, Crick construed a simple, transitional DNA strand which was supposed to be an intermediate form between RNA and the contemporary DNA strand. They proposed that in the erstwhile DNA molecule, there were two complementary chains formed out of 2 bases (instead of 4 in contemporary structure) which are parallel, rather than antiparallel (Woese, 1967). In such a model, those two complementary bases were predicted to be A: U (or T) or G:C whereas the biochemist, Leslie Orgel, in his correspondence observed that in the 2 base pair model, complementary base pairing in the primitive, parallel DNA thread, was found to be A: I (Inosine). Crick (1968) indirectly endorsed the possibility of structural formation of primitive nucleic acid, when he said, "*It is not certain that a double helix can be formed having a random sequence of A's and I's on one chain and the complementary sequence (dictated by A-I and I-A pairs) on the other chain, but it is not improbable, especially as the RNA polymers poly A and poly I can form a double helix.*" Regarding the evolution of A and I, Crick (1968) and his associates suggested that "Adenine" was the most common base in the "pre-biotic soup" and the "Inosine" was the derived one (the delaminated form of Adenine base). He further argued in favour of the evolution of primitive form nucleic acid stating, "*If we can use the present code as a guide (though we shall argue later that this may be misleading), it is noticeable that the triplets containing only A's or G's in their first two bases (the bottom right-hand corner of the Table) [contemporary genetic code] do indeed code for some of the more obviously primitive amino acids.*" Then a series of composite changes in the evolution of base pairings transferred the primitive structural pattern of nucleic acids was observed, gradual changes of pairing from A:I, to the contemporary form such as A:T. The gradual mutations yielded Uracils and Cytosines that took part in complementary base pairing process in place of the old base pairings, where replicase enzyme (which is considered as the RNA transformed structural entity) helped to expedited all these base-pairings; resulting in formation of new codons on its way to the evolution of biosynthesis of protein. It was further assumed that most likely I (Inosine) was gradually substituted with G (Guanine) base. So, Crick (1968), his

colleagues and contemporary molecular biologists finally came to the
conclusion that the idea and observation of the early molecular biologists
which recognized the existence of primitive nucleic acid (with two bases)
was credible on the evolutionary route and shifted from evolution of
primitive nucleic acid to the diversification of advanced form of nucleic
acid. Though, factors like changes in base pairs, formation of new codons,
changes in old codons, change of structural configuration of nucleic acids
and polymerization of biosynthesis etc. would influence important genetic
principles like fundamental structural formation of the genetic code, there
were no major visible changes in the structural or functional changes of
proteins. Such changes could alter the concept of species and extent of
multitude of biological evolution which started with a simple, unicellular,
microscopic organism and evolved in to the diversified multicellular,
advanced organism of today. Yet, the contemporary genetic codes,
supposedly, have gone through different phases of persuasive change where
universality has been compromised to some extent in the course of time and
space.

Since the beginning of '90s, the advancement of molecular biology have
publicized contemporary investigations in such a way that it has often been
claimed erroneously that "genetic code of some organism X has been
cracked by genomics," which is difficult to swallow as genetic code is not a
structurally and functionally unique nucleotide. Rather, its expression would
be available in the form of protein translation (Keeling, 2016). The genetic
code should be regarded as a set of rules or guidelines (as well as 64 codons)
that help to interpret molecular language of nucleotides with 20 amino acids
which help in the synthesis of diverse array of proteins. Apparently, it is
difficult to correlate the evolution of genetic code to the evolution and
diversification of life due to its "highly conserved," non-diversifying,
universal nature of recognition, witnessed among different group of
organisms. Yet, the landslide progress in contemporary molecular biology
ascertained that "some variants do exists," which made molecular biologists
use the term "Canonical nature of genetic code" rather than "Universal
nature of genetic code" as the identity of the "non-canonical code." Its
integration to the protein translation systems prompted the molecular
geneticists to trace the distribution in "Tree of life" , the ultimate
phylogenetic model of diverse array of organisms on earth (Keeling, 2016;
Zahonova et al, 2016). Apparently, genetic code is a fine blend of perception

and concept that could capture the imagination of the most intelligent molecular biologists who started their journey in early half of the 20th century when they tried to interpret the 4-lettered language of DNA to encode 20 letters of protein (Gamow, 1954a). Genetic code should not be considered as the best discovery as that would be idealistic rather it was considered as "the series of contingencies" which has gone through the perception of Crick (1968) called "Frozen Accident." He preferred to see it as the Universal/conservative model, having little scope for change, followed by the findings that the primitive genetic code evolved with fewer number of codes, having the potential to translate only fewer number of amino acids. That level of perception of the existence of primitive code helped to understand about its evolution and diversification in numbers and its potential to translate proteins with advanced structure and functions those made–up of more amino acids. So, ideas of evolution of the genetic code in terms of structural orientation and function led to the review of its universal status and justifiably consider it as the canonical model instead (Keeling, 2016). However, it was found that before the LUCA or Last Universal Common Ancestor of all living descendants, some sort of "Powerful selective Constraints" kept the Genetic code frozen in almost all genomes which got diversified in course of evolution (Keeling, 2016).

The contemporary studies in the molecular footprint of prokaryotic genomes, along with analysis of the phylogenetic distributional patterns of non-canonical codes have revealed that in the lower group of organisms universal code was hardly used in protein translation process and this evidence showed in the comparative molecular distinctive study between mitochondrial genome and nuclear genome of the prokaryotic organisms. Furthermore, Keeling, (2016) has stated, "*the vast majority of changes to the code we presently know of involve termination or stop codons being reassigned to encode an amino acid. This may be a true reflection of natural diversity of the code because stop codons are by definition rare (only one of three possibilities appearing per gene, whereas even rare amino acids are typically found many times) and the fidelity of termination is potentially less critical then other possible changes.*" As the codons act as interface to translate the sequence of nucleotides to a sequence of amino acids, if gene or amino acids any of these flanks of equilibrium is unknown, genetic code is no longer a principle anymore. Keeling, (2016) considered the stop codons as "sore thumb" DNA such that it disrupted the coding region, which is difficult to overlook, and such codes generated from stop codons could be inferred from DNA sequence. Whereas, the molecular biologists observed that in a code where "two amino-acid encoding codons were altered," the DNA sequence appeared in slightly distinct form, rather than translated by a distinct genetic code (Keeling, 2016). While working on the non-canonical nuclear genetic code with three terminal codons (like UAA, UAG, and UGA) of *Blastocrithidia*, Zahonova et al (2016) surprisingly noticed, that UAA and UAG UAA was found to play dual action as they are able to encode amino acid glutamate along with stereotypic task that act as stop codon. Since the molecular investigations by Schulz and Yarus (1994), particularly on mutational effect on tRNA, it was observed that the mutational effect might alter the protein translational potential of codons and they might evolve with unique protein translation potential.

According to the contemporary concept of the genetic code regarding its universality and persistence, it is difficult to accommodate any ideas about structural modification of genetic code. As Crick (1968) opined, "*A change in codon size necessarily makes nonsense of all previous messages and would almost certainly be lethal. This is quite different from the idea that the primitive code was a triplet code (in the sense that the reading mechanism moved along three bases at each step) but that only, say, the first two bases were read. This is not at all implausible.*" Regarding solicitation in favor of primitive codes, the biochemists, molecular geneticists and evolutionary biologists argued that most probably primitive codes dealt with fewer amino acids as the diversified/advanced group of organisms did not exist at that time. Hence, biosynthesis of complex-structured proteins (and their building blocks, the amino acids) were less likely to exist as amino acids such as "tryptophan," "methionine" etc. evolved much later than their primitive counterparts of "alanine,' "aspartic acid," "glycine," "serine" etc. (Crick, 1968). Jukes (1973), in his meticulous study on the evolution of primitive genetic code stated that there were less than 20 amino acids that evolved during evolution of primitive genetic codes and suggested that primitive amino acids like arginine were functionally replaced by ornithine during the diversification of protein synthesis process in the course of time. In 2010, identical observations were made by the eminent Japanese molecular biologist, Dr. Shin Chi Yokobori, from Tokyo University of Pharmacy and Life Science, along with co-investigator, Dr. Takuya Ueda, from National Institute of Advanced Industrial Science and Technology, Tokyo, Japan who proposed that the ancestral genetic code started with few of amino acids and evolved to the universal form used by most living and advanced group of organisms. Yokobori et al (2010) further observed that factors like genome economization, directional mutation pressure on genome and evolution of tRNA moulded the evolution of primitive genetic code and eventually changed and evolved to be the contemporary universal form. The meticulous studies of Yokobori et al (2010) showed that mitochondrial genetic code of the different group metazoan lineages (diverse group of multicellular organisms) varied in different groups of organisms and remarkably distinct from the universal genetic code. It rendered the idea for changes of genetic code that had gone through the course of time and evolved in diverse group of organisms with structural advancement. Hence, universal constancy of genetic code as a model perception was compromised in course of advancement in molecular biology. In support of the evolution of primitive codons, Jukes (1973) referred to some amino acids as primitive such as alanine, glycine, serine etc. having a minimum of 4 codons or more and the sequential arrangement of triplet base followed according to the quartets, (U--> C ---> A ---> G) such as: GCU, GCC, GCA, and GCG. The four codons encoding alanine and the first two letters of the triplet specified the particular amino acid. So, Jukes (1973) construed that "quartets" of codons would be restricted to encode around 16 amino acids. In his hypothesis, Jukes (1973) further elucidated that primitive genetic codes were able to code 10 amino acids, out of 10, 6 amino acids were encoded by 8 codons and remaining 4 amino acids were encoded by 4 codons and those 10 amino acids could be part of protein synthesis of primitive codon on the earth recognized predominantly as the RNA world. It was further extrapolated by Jukes (1973) that expansion of encoding pattern of 18 amino acids from erstwhile amino acids evolved as a result of mutation of primitive amino acids parallel to modifications of tRNA in such a way that G: U or U: G was suppressed. The gradual increase of amino acids from 10 to 15 and from 15 to 18 and finally to 20 led to better synthesis of protein through the process of evolution and diversification by offsetting the role of nucleic acid as genetic material. It rendered structural and functional modifications to tRNA and further expansion of genetic code as a process of modification of primitive code to evolve as contemporary genetic code—which is 64 in numbe—and could efficiently encode 20 amino acids to synthesize proteins of diverse structural and functional orientation to support evolution of an array of organisms on Earth (Sharma, 2006). Scientists suggested further that though the number of amino acids supposed to exist during the evolution of primitive codes was limited, the existence of non-sense codons was less likely, as, expression of messages would have a considerable number of gaps. There are a number of interpretations and ideas to support these hypotheses and one of the popular interpretation was the recognition of one base of the triplet codon. Crick (1968) and Woese (1965) fostered the recognition of the middle base of the triplet e.g. recognition of middle base "A," stands for an acidic amino acid and "U" stands for any hydrophobic amino acid etc. Though Crick (1968) was in favor of primitive genetic codes constituting two bases, rather than present day genetic code constituting four bases. According to the contemporary progress in molecular biology, particularly, the evolution of genetic code in relation to biosynthesis of protein as well as defining the hereditary transmission of molecular footprint of life or 64 (triplets) to 20 (amino acids) was one of the best evolutionary gift in the history of biological evolution. The contemporary genetic code was found to be one of the best genetic immunity as well as safeguard for the cell to protect it from the lethal effects of substitution mutations. The contemporary genetic code evolved marvellously and sustained so that even a single base difference in triplet codons did not stop it encoding from the same amino acid or an amino belonging to an identical chemical group (Smith, 2008). So, the modern genetic code has a close analogy with a program of in-built firewall protection so it could nullify the possibility of any error that might take place during translation, particularly during complementary coupling of codon and anticodons on the way to protein translation as a result of rejection of the incorrect duplex by the concerned ribosomes (Lim and Curran, 2001). As an example, the four codons of the amino acid 'valine' was considered and out of the 4 codons, GUC was considered here. The occurrence of substitution mutation at 3' position of GUC could change the C to A or U or G resulting in change of the codon GUC to GUA or GUU or GUG and these newly altered codons encode the amino acid 'valine.' In another scenario, if G at 5' position were to be substituted with A, the new codon appeared with the triplet AUC. AUC codon could encode the amino acid Isoleucine which is functionally and structurally identical to valine. Hence, scientists came to the conclusion that the modern genetic code evolved in such way that it was competent to minimize ORF (Open Reading Frame) degradation (Brown, 2002). It was observed that the modern genetic code was not absolute universal stop codons like UAA and UAG often went through frame-shift mutations to be recognized as stop codons while having the potential to excel in dual mode as such as to encode glutamate or aspartate. The above observation led contemporary scientists (Itzkovitz and Alon, 2007) to believe that in stereotypic genetic code, translational correction of any frame-shift error was paused or was much faster (99.3%) than in alternative codes. Going a step forward, Itzkovitz and Alon (2007) expressed their concern at the future potential of the contemporary genetic code in shaping biological evolution and stated, *"We find that the universal genetic code can allow arbitrary sequences of nucleotides within coding regions much better than the vast majority of other possible genetic codes. We further find that the ability to support parallel codes is strongly correlated with an additional property — minimization of the effects of frameshift translation errors."*

References

Alberts, B., Johnson, A., Lewis, J., Raff, M., Roberts, K., and Walter, P. (2008) Molecular Biology of The Cell, 5th edition, New York: Garland Science.

Brenner, S., Stretton, A.O.W. and Kaplan, S. (1965). Genetic code: The 'nonsense' triplets for chain termination and their suppression. Nature 206:994-998.

Brown, T. A. (2002). Genomes. 2nd edition. Oxford: Wiley-Liss; Available from:
https://www.ncbi.nlm.nih.gov/books/NBK21128/

Crick, F.H.C. (1958). On protein synthesis. *Symp. Soc. Exp. Biol.* 12: 138–1163.

Crick, F. H.C. (1966). Codon--anticodon pairing: the wobble hypothesis. J Mol Biol. 19(2):548-55. doi: 10.1016/s0022-2836(66)80022-0.

Crick, F.H.C. (1968). The origin of the genetic code. Journal of Molecular Biology. 38 (3): 367–79. doi:10.1016/0022-2836(68)90392-6

Gamow, G. (1954a). Possible Relation between Deoxyribonucleic Acid and Protein Structures. Nature. 173:318.

Gamow, G. (1954b). 'Possible Mathematical Relation between Deoxyribonucleic Acid and Proteins' Det Kongelige Danske Videnskabernes Selskab, Biologiske Meddelelser. 22:3–13.

Gamow, G. (1954c). Letter to Watson, 7 March 1954, Wellcome Library, PP/CRI/D/4/1: Box 29,
<http://library.wellcome.ac.uk/player/b18180279>

Gamow, G. (1954d). Letter to Crick, 8 March 1954, Wellcome Library, PP/CRI/ H/1/42/5: Box 72,
< http://library.wellcome.ac.uk/player/b18175089>

Gamow, G. (1954e). Letter to Crick, 27 May 1954, Wellcome Library, PP/CRI/H/1/42/5: Box 72,
< http://library.wellcome.ac.uk/player/ b18175089>.

Goodman, H. M., Abelson, J., Landy, A., Brenner, S., and Smith, J. D. (1968). Amber suppression: a nucleotide change in the anticodon of a tyrosine transfer RNA. Nature. 217(5133):1019-24. doi: 10.1038/2171019a0.

Itzkovitz, S., and Alon, U. (2007). The genetic code is nearly optimal for allowing additional information within protein-coding sequences. Genome Res. 17(4):405-12. doi: 10.1101/gr.5987307.

Jukes, T. H. (1973). Possibilities for the evolution of the genetic code from a preceding form. Nature. 246(5427): 22–26.

Keeling, P. J. (2016). Genomics: Evolution of the Genetic Code. Current Biology 26(18):R851-R853. DOI: 10.1016/j.cub.2016.08.005.

Khorana, H.G. (12th December, 1968). Nucleic acid synthesis in the study of the genetic code. Nobel lecture.
https://www.nobelprize.org /uploads/ 2018/06/khorana-lecture.pdf

Khorana, H.G. (1979). Total synthesis of a gene. Science 203(4381): 614-625.

Lim, V. I., and Curran, J. F. (2001). Analysis of codon: anticodon interactions within the ribosome provides new insights into codon reading and the genetic code structure. RNA. 7(7): 942-57. doi:10.1017/ s1355838201 00214x.

Madison, J. T., Everett, G.A. and Kung, H. (1966). Nucleotide Sequence of a Yeast Tyrosine Transfer RNA. Science 153(3735): 531-534.

Nirenberg, M. W., and Matthaei, J. H. (1961). The dependence of cell-free protein synthesis in *E. coli* upon naturally occurring or synthetic polyribonucleotides. Proc Natl Acad Sci U S A. 15;47(10):1588-602. doi: 10.1073/pnas.47.10.1588.

Pelc, S. R., and Welton, M. G. (1966). Stereochemical relationship between coding triplets and amino-acids. Nature. 209(5026):868-70. doi: 10.1038/209868a0.

Schultz, D. W. and Yarus, M. (1994). Transfer RNA mutation and the malleability of the genetic code. J Mol Biol. 235(5): 1377-80. doi: 10.1006/jmbi.1994.1094.

Sharma, A.K. (2006). Plant genome: Biodiversity and Evolution; Phanerogams (Angiosperms-Dicotyledons). Vol 1 (part-c). CRC Press. P. 595.

Smith, A. (2008) Nucleic acids to amino acids: DNA specifies protein. *Nature Education* 1(1):126.

Tsuchiya, H., Sawamura, T., Harashima, H., and Kamiya, H. (2005). Correction of frameshift mutations with single-stranded and double-stranded DNA fragments prepared from phagemid/plasmid DNAs. Biol Pharm Bull. 28(10):1958-62. doi: 10.1248/bpb.28.1958.

Woese, C. R. (1967). The genetic code: The molecular basis for genetic expression. Harper & Row, New York, p.200.

Yokobori, S-I., Ueda, T. and Watanabe, K. (2010). Evolution of the genetic code. John Wiley & Sons, Ltd.
https:// doi.org/10.1002/ 9780470015902.a0000548.pub2

Zahonova, K., Kostygov, A. Y., Sevcikova, T., Yurchenko, V., and Elias, M. (2016). An Unprecedented Non-canonical Nuclear Genetic Code with All Three Termination Codons Reassigned as Sense Codons. Curr. Biol 26(17): 2364-2369.

Chapter 4

Is It Emerging from Shadows or Plunging in Darkness of Evolution? Peeping Through the Specs of Gene Mutations and Modulation of Gene Expression

All multi-cellular life on earth is undergoing inexorable genome decay because deleterious mutation rates are so high...and natural selection is so ineffective in removing the damage....like rust eating away the steel in a bridge, mutations are eating away our genomes and there is nothing we can do to stop them.

---- Alex Williams

The intensive genomic study confirms that all cells of an organism possesses the same genome. The presence of such "consistent omnipresence" of genome indicate existence of a firm regulatory system that controls gene expression. Digging dipper into the molecular regulatory pathways, scientists found that there are two sites of regulation for genic functions and two separate mechanism to control gene activity, independently. Out of the two, one mechanism is operated from cytoplasm and it is thought to be regulating synthesis and functional aspect of an enzyme. The second one operates from the nucleus which regulates the action of structural genes (Pandey, 2017). Modulation of gene expression is further described under two stages-

a. Modulation of gene activity in the cytoplasm:

An example of the enzyme action of tryptophan synthetase on the growth of *E.coli* may be stated, where its application in the culture medium of *E.coli* slows down the growth of *E. coli* and is referred to as "enzyme repression." This experiment has been described on the basis of the end product which inhibits the enzyme synthesis involved in enzyme-substrate reaction (Stadtman et al, 1961). The case of inhibition of enzyme action in isoleucine synthetase where the accumulation of the end product leads to the synthesis of the substance has been recognized as "feed-back inhibition" (Stadtman et al, 1961).

b. Modulation of gene activity in the nucleus:

The genes at a cellular level expresses a number of functions. The studies on the synthesis of β-galactosidase enzyme in *E.coli* concluded that all genes do not undertake synthesis of enzyme proteins (Jacob and Monod, 1962). Henceforth, Jacob and Monod (1962) promulgated that there must be a group of genes responsible for the regulation of action of structural genes (the structural gene code for biosynthesis of a wide array of cellular proteins) and these supervising genes act on the structural genes and are recognized as "regulator genes." The primary function of the regulator gene is to determine whether the structural gene or a group of linked structural genes would be transcribed to mRNA (which would undertake translation of proteins) in a given period of time.

Hence, Jacob and Monod (1962) proposed a unique model to define the modulation of gene expression called "Operon Model.' They observed that if the enzyme β-galactosidase is synthesized by *E. coli* then this enzyme catalyzes lactose (which is a disaccharide) yielding monosaccharides: glucose and galactose. So, in the absence of lactose substrate, synthesis of β-galactosidase enzyme is not a necessary task at the gene level. On the basis of the results mentioned above, scientists inferred that either physiological, metabolic or environmental conditions—individually or in the composite form—modulate the expression of genes. Further, they interpolated that the differentiation of structural and functional form of any organism from juvenile to matured state is a result of interactive, complex factors, modulating the expression of a number of genes. In the Operon model, lactose is recognized as "inducer" and to check the functional potential of the inducer, the induction of "repressor" has been proposed and tested. It was further observed that the actions of the repressor is reciprocally controlled by the synthesis of the inducer as it could remain inactive in the presence of a co-repressor (Stryer, 1988). So, the Operon model clearly elucidates two different systems to control gene expression, which are as follows:

a.   Inducible Systems: It involves the activation of repressor in the presence of inducer which then becomes an inactive repressor. The inactivation of the repressors leads the structural gene functional to synthesize mRNAs and produce proteins (enzymes) to regulate metabolic activities.

b.   Repressible Systems: It involves activation of repressors due to the presence of a co-repressor. As long as repressors remain active, the actions of the structural genes are on hold stalling the biosynthesis of mRNA.

However, "transcriptionally regulated system" of the Lactose Operon model, elucidated by Jacob and Monod (1962), portrayed the modulation of polycistronic structural gene which is regulated by regular genes along with a common promoter. This kind of model is very common in bacterial organisms and referred to as "Operon" e.g. *ara* operon, *lac* operon, *val* operon etc.

The lac operon constituted of one regulator gene, *i* gene (which was recognized as inhibitor, rather than inducer) and three structural genes: z gene, y gene and a gene. Functions of the constituent genes of the lac operon model (Stryer, 1988) are as follows:

i gene → encodes the repressor of the lac operon.

z gene → encodes the β-galactosidase protein (enzyme). It catalyzes the hydrolysis of lactose (disaccharide) into two monosaccharides: galactose and glucose.

y gene → encodes the permease protein (enzyme). This enzymes enhance the extent of permeability of the cell to the β- galactosides.

a gene → encodes the transacetylase protein (enzyme). Hence the three structural gene products of lac operon model, which comprises z gene, y gene and a gene is involved in the metabolism of lactose. In other operon models like *trp* operon or *ara* operon, the constituent structural genes follow the same metabolic pathway to metabolize tryptophan (amino acid) or arabinose (sugar) substrate. The diagrammatic sketch of the lac operon model is presented in fig.4.1.

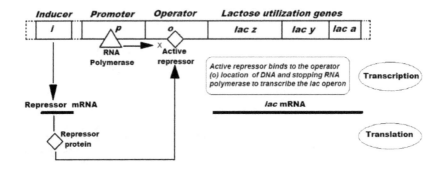

Type A. No synthesis of Lactose utilization proteins in absence of inducer

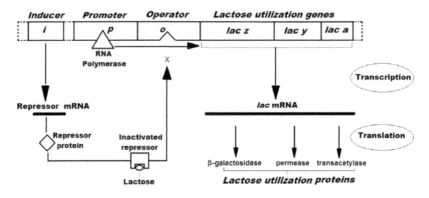

Type B. Synthesis of Lactose utilization proteins in presence of inducer

Fig. 4.1. The diagrammatic outline of *lac* Operon model.

In the lac operon model, the β-galactosidase enzyme acts upon the substrate lactose which modulates turning on and off of the operon, hence lactose is recognized as an inducer. Instead of a conventional source of sugar, like glucose, if lactose is used as a source of sugar in the culture medium for bacterial growth, then the transportation of lactose through the permeable membrane of the bacterial cell membrane with the help of permease enzyme play a cardinal role. The lactose sugar coordinates the function of Operon in the following way (Stryer, 1988). The repressor protein of the Operon is synthesized from the regulator gene, known as *i* gene. The repressor protein firmly adheres to the operator region (o) of the Operon and restrains the RNA polymerase enzymes from transcribing operon. In the presence of lactose (which acts as inducer), the repressor is functionally inactivated as a result of repressor-inducer interaction. The inactivation of repressor enables the RNA polymerase enzyme to reach the promoter region and expedite the transcription process. The regulation of gene expression, specifically the modulation of Lac Operon, by the repressor is popularly referred to as "negative transcription regulation" (Shaw, 2008). The molecular geneticists have meticulously studied the impact of mutation on the regulator gene "i" and the operator gene of the Lac Operon model. The critical review of the scientists revealed that formation of repressor is blocked by mutational effect on the gene and in absence of formation of repressor, the uninterrupted biosynthesis of enzyme continues (Stryer, 1988). The concerned bacterial strains related to this property is referred to as "constitutive mutants" or "constitutive strains." The molecular geneticists further observed that mutational impact on the operator gene restrained the repressor to coordinate its function in an attempt to control the operator site of the Lac Operon. On the basis of the functional mode and the nature of mutational impact, either on the regulator zone or on the operator zone of Lac Operon model of *E.coli*, constitutive mutants are broadly categorized as either "regulator constitutive" or "operator constitutive" (Fiethen and Starlinger, 1970; Lodish et al, 2000). The diagrammatic outline of (a) regulator constitutive and (b) operator constitutive are presented in fig. 4.2.

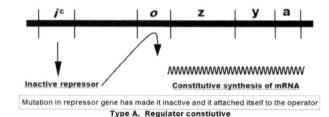

(Left) Fig. 4.2. The diagrammatic sketch of Regulator constitutive and Operator constitutive.

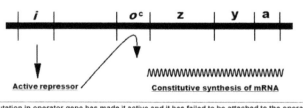

The recent advancement in the field of molecular biology and genomics unraveled the mutation BRCA, i.e. mutation in BRCA1 and BRCA2 genes that helped women make decisions about career planning and a personal life plan regarding relationships, conception, child-birth and treatment etc. Clinical pathologists and medical oncologists reveal that mutations of BRCA1 and BRCA2 genes lead to diseases like hereditary breast and ovarian cancer syndrome in women who are carriers. (Weitzel et al, 2007). The level of social awareness (among men and women) about the root cause of cancer not only helps to understand the cause and effect of these diseases, but also help them review the social, behavioural and psychological consequences of cancer and impact of mutations in human systems (Leroi, 2003). So, in the epoch of Anthropocene, mutation should not to be considered as a biological process, rather, as genetic alterations. It is an alteration of our "blue print of life," the DNA profile of living organism, which encodes the phenotypic expression of the concerned living organism.

Though it is an utopian idea to find any biological organism as well as a cellular entity without any impact of mutation in its natural condition, in experimental environment or in laboratory condition, if any organism is traced without any impact of mutation, that organism needs to be recognized as "wild"; otherwise almost every organism would be a "mutant." Mutation, the sustainable change of the sequence of nucleotides or genome or extra-chromosomal DNA or any other genetic constituent of an organism or virus, leads to the alteration of genotype of the concerned organism which may or may not trigger a substantial phenotypic change but leads to alteration of a number of important biological mechanisms of life such as evolution, carcinogenesis, immune systems etc. As mutation directly interferes with the nucleotide sequence of any organism, it alters the structural and functional properties of genic and non-genic regions leading to the alteration of a protein produced by the organism in its normal state of living. A contemporary experiment on the mutation-induced genetic changes of different species of *Drosophila*, by a group of molecular biologists, revealed that mutagenesis led to changes in the configuration of proteins, where 70% of the amino acid polymorphisms have produced harmful or lethal effect, and remaining changes have been found to be either neutral or partially beneficial for evolution (Sawyer et al, 2007). Investigations on the molecular pathways of cancer revealed that most of the biological organisms have inherent counter-active mechanisms against mutation that help to repair the mutagen-induced damages and restore the mutated DNA sequences to its previous state (Bertram, 2000). Apparently, in most occasions, the mutational changes of gene and its change in the functional and configurational pattern translated to proteins, would not achieve sustainability as it is found in the most altered proteins. This is because the phenotypic expressions are harmful and would not comply with the fitness of natural selection. Hence, mutation is to be found harmful for those that do not go with natural selection. This part of mutation is quite enigmatic as evolutionary biologists have thousands of hands-on evidence to prove how mutation could help create new traits or species. As an example, evolution of a number of species of "Abalone shellfish" could be considered, where mutation of the protein (key) on the surface of its sperms expedite its binding to the surface of the egg (lock) as in normal situation. The possibility of the fertilization of the eggs of a number of species of "Abalone shellfish" was found to be an impossible task unless the mutated sperm became biochemically as well as structurally compatible with the egg to expedite fertilization (LePage and de Perre P, 2012). Based on the apparent success of biological evolution, driven by mutational evidence, mutation was found to be an auxillary biological mechanism that favours natural selection. Now, let's walk through the lessen known evolutionary trail of mutational evidence to find out whether mutation got any support in fitness attributed by natural selection or has it always played an antagonistic role of nipping the bud of lethal evolutionary potential.

Evolutionary biologists reviewed the historical perspective of drinking milk of the domesticated cattle by the human kids and revealed that the milk digestion potential of the kids stalled when most of the kids became teenagers and stop drinking the cattle milk to rely on other food sources for their sustenance. The meticulous observation of reviewing this milk drinking issue by researchers on a number of adults revealed that as an effect of domestication of cattle, a number of human populations in Africa and Europe, independently acquired the mutation ability to digest milk, even after attaining adulthood (LePage and de Perre P, 2012). Population geneticists ascertained that genetical success of such mutational incident was strongly supported by natural selection. The evolutionists considered that the change of bio-social environment of human population, like the domestication of cattle (cows and buffaloes) changed the availability of their food resources, ultimately triggering the genetic mutation of the adults to cope with drinking milk in adulthood. However, creationists put forth their counter argument that loss of the molecular switch of the milk-digesting enzyme in the adulthood is responsible for gaining the milk digesting abilities of the adults (LePage and de Perre P, 2012).

So, gene mutations could be defined as permanent changes in DNA sequence which constitutes gene and as a result of mutation the altered sequence of DNA is traced in the genetic profile of the concerned person and is found to be different from most people, having unaltered DNA sequences as well as non-mutated gene. The extent of mutation varies from a single base-pair alteration to changes in a fairly larger segment of chromosome having multiple genes. So gene mutations could broadly be categorized into two groups (Griffiths et al, 2000; U.S. Department of Health and Human Service, 2009):

a. Germline mutations or Hereditary mutations: The offsprings inheriting this kind of mutations have mutated DNA from their male and female parent which is transferred through the sperm and egg to the embryo. The mutated DNA acquired through germs cells from their parents is found in every cell of the offspring.

b. Somatic mutations or Acquired mutations: This kind of mutation is witnessed in certain phases of its life cycle of any organism and it has been observed in some somatic cells of the body of the concerned organism. This kind of mutation is not acquired through germ cells, so it is not inherited through generations. This kind of mutation is found to occur as a result of physio-chemical mutagens (e.g. UV radiations, gamma radiations) or it might happen at gene level as a result of formation of erroneous copies of DNA during replication.

The scientists observed that genetic mutations which is recognized as "de novo mutation" might be somatic or hereditary in nature (Samocha et al, 2014). The expression of de novo mutation is witnessed in offsprings born with genetic disorders, while, there may not have been any early record of genetic disorders in its parents or family history. It may be concluded that either the parental germ cells are mutant which gave rise to a mutant embryo or the fusion of normal germ cells gave rise to an embryo which mutated immediately after embryo formation. Whatever the way of fertilization embryo gave rise to mutant offspring with all mutant body cells (U.S. Department of Health and Human Service, 2009). On the other hand, the somatic mutation happens in a single cell in the early stages of embryonic development and is called mosaicism. It was observed that mutation induced mosaicism might or might not cause any disease in the offspring (Yandell, 2014). When genetic alterations were observed in more than 1% of the population, they were recognized as polymorphisms. These would either cause physical disorders or simple phenotypic variations without any health issues (U.S. Department of Health and Human Service, 2009).

The evolutionary biologists noticed that genetic recombination and mutations led to genetic variations which regulates the extent of evolutionary process of any organism's population dynamics over generations. The genetic variations found to change gene actions, protein synthesis or enzyme actions, ultimately lead to unique traits in an organism. When a trait is found to be useful for that organism, it is supposed to achieve sustainability. They survive, reproduce and pass the favored genetic variation to the next generation. The new traits persist in the following generations as it goes through the natural selection and is passed down to most individuals in the new population. This results in the formation of new population which is distinct from its ancestral lineage and often forms different varieties or species.

So, the next point of concern is whether any mutation leads to the emergence of new traits and evolution of new species or not. Evolutionary biologists observed that it's only hereditary mutations induced by genetic variations that is passed through the parental lineage by means of germs cells (through fertilization of sperm and egg) (Griffiths et al, 2000). Whereas, the acquired mutations or mutations of somatic cells are not passed from one generation to the next and no unique trait is formed or selected through natural selection process. Hence, such somatic mutations have neither impacted evolution nor have been proven helpful or harmful for derived populations in terms of gene actions or protein (enzyme) synthesis (Griffiths et al, 2000). However, any mutational changes or emergence of unique traits found to be beneficial for a group of organisms, living in a particular environment might be lethal for a different population surviving in different environmental condition. Mutation creates an antibiotic resistant bacteria which could survive in a particular media constituting high content of antibiotic "X" but it would not be able to survive in the media constituting of antibiotic "Y." It was predicted that the genetic disease, persisting in population generation after generation was supposed to be weeded out by natural selection, but that did not happen in reality. The explanation is not very straight forward for this answer. For sickle cell anemia disease, two copies of mutated *HBB* gene in each cell causes the disease whereas, one copy of gene was found to be forming resistance against malaria (Carlice-Dos-Reis et al, 2017). For Huntingdon disease, mutation induced gene variation and expression of disease is not witnessed until child birth as harmful traits passed from parent to the offspring have a delayed expression, called "reduced penetrance."

However, the DNA sequence of a gene found to be altered, results in changes in the gene structure, protein configurations and functions in a number of ways. The main categories of gene mutations (Silver et al, 1953; Jones et al, 1998; Griffiths et al, 2000; Paulson, 2018) are as follows:

A. Insertion: In the insertion related mutational changes, a number of extra base-pairs inserted in the DNA sequence alter the existing DNA sequence of the gene. It leads to structural and functional changes in the protein translation encoded by altered DNA sequence of the mutated gene.

e.g. ATTGGAC (Original base-pair sequence)

ATTGGTTGAC (Insertion of TTG in the mutated DNA sequence)

B. Deletion: In this type of mutation a number of base-pairs of a gene or the entire gene or a number of genes in a sequence is deleted. The deleted DNA, supposedly alters the structure and function of protein encoded by mutated gene.

e.g. ATTGGAC (Original base-pair sequence)

ATTAC (Deletion of GG in the mutated DNA sequence)

C. Duplication: In this type of mutation, a duplicated segment of DNA is constitutes mutated DNA, which might be copied erroneously or more than once, resulting in alteration of structure and function of proteins encoded by mutated DNA.

D. Substitution: In this kind of mutation, a particular base switches position with substitute base in the DNA sequence of a gene leading to three possible scenarios-

i. The changing of a codon which encodes the same amino acid (having more than one triplet code) so there is no change in the structure and function of protein in post-mutation phase and is recognized as "silent mutation" or "neutral mutation." It's difficult to distinguish the organism of "silent mutant" type and "wild" type as both are phenotypically identical.

ii. The alteration of a codon which encodes a different amino acid, leading to structural and functional alteration of proteins is synthesized by mutated sequence of DNA. For example, the occurrence of genetic disease like Sickle Cell Anemia is a result of substitution in the beta-hemoglobin gene. It changes a single amino acid in the protein leading to altered identity and function in the protein encoded by mutated DNA sequences.

iii. The changing of an amino-acid-coding codon to a single "stop" codon is a shift in a triplet base sequence, resulting in the formation of an incomplete protein. For example, stop codons like TAG, TAA and TGA act on DNA molecule wile stop codons like UAG, UAA and UGA act on RNA molecule. Such formation of incomplete protein could jeopardize the existence of an organism as the function of structurally incomplete protein will not perform any pre-destined task.

E. Point Mutation: The occurrence of this kind of mutation is the result of the replacement of one nucleotide by another in a particular sequence of a gene. It causes negligible extent of phenotypic changes and is further categorized in two groups:

i. Translation: In this category of point mutation, a purine base is replaced by another purine base or a pyrimidine base is switched with another pyrimidine base.

ii. Transversion: In this category of point mutation a pyrimidine base is switched with a purine base or vice versa.

F. Frameshift mutation: The occurrence of frameshift mutation occurs as a result of alteration of the reading frame of a gene, caused by addition or deletion of the base changes in the DNA sequence. The reading frame of the gene consists of three bases in sequence—forming triplet codons—and each triplet codon encodes an amino acid. The alteration of bases like insertion, duplication or deletion would alter the sequence of triplet codons, resulting in structural alteration of codons, resulting in incompatibility of encoding a particular amino acid and sequential alteration of amino acids leading to the formation of 'non-functional' protein. e.g.   THE
     CAT   EAT   RAT (Original reading frame)

THE CAT EAT RAT (Deletion induced change, causing alteration of reading frames)

HEC   ATE   ATR   AT (Altered reading frame yielding non-functional protein)

On the basis of transcriptional property, point mutation is further studied under Missense mutation and Nonsense mutation.

G. Missense mutation: In this type of mutation, there is a change in one DNA base-pair which leads to the alteration of the gene, changes in one amino acid for another and in the configuration of the protein made by the mutated gene. When the altered amino acid nesting in the active site of protein forms non-functional protein, it causes lesser extent of phenotypic change of the mutant organism.

H. Nonsense mutation: In this type of mutation, there is a change in one DNA base-pair but it does not substitute one amino acid for another. Instead, it signals the cell to stop the completion of protein translation. So, this type of mutation leads to the formation of incomplete protein molecule whose functions remain ambiguous. Nonsense mutation causes greater extent of phenotypic changes in the mutant organism.

I. Repeat expansion: A repeat expansion is a particular type of mutation that increases the number of repetition of short DNA sequence. The nucleotide repeats are short DNA sequences that are repeated a number of times. For example, trinucleotide repeats are made up of three base-pair sequences, likewise, tetra nucleotide repeats are made up of four base pair sequences. However, this repeat expansion mutation to improper protein function.

The widespread functional impact of gene mutations as a result of changes in DNA sequences alters the structural and functional identities of any organism at a cellular level due to variations in proteins. Such changes might be carried forward through the generations or might be restricted to a particular generation without having much impact in steering the mode of evolution. Whether the mutation has the potential to cause any disease in an organism by somatic mutation or germ-cell mutation, the changes would lead further natural variations in extant populations which would help to evolve a new species. The reversal of the alteration of mutational changes could be a possibility of reverse mutation and present the diversified functional potential of mutations. Mutation could be a continuous biological process which is either integrated to the internal physio-chemical actions at the molecular level in any cellular organism, or it might be triggered by the environmental surroundings of the organism. As we know that at the time of cell division DNA is replicated and any erroneous copying would create a new DNA deviating from its archetypal template, thus leading to mutation. In other way, the exposition of any organism in any environment, which is full of physical mutagens (like UV rays, gamma rays, high temperature etc.) that damage/alter the DNA of the concerned organism at a cellular level and break it down causes mutation. Though such breakaway parts of DNA go through inherent, genetical, self-repairing process done by the internal enzymes but such genetical retrofitting DNA repair lead to slightly different DNA strands from the original one, thereby, causing mutations.

From a diverse array of mutational changes of genes, it seems that any mutation could create a genetic disorder but clinical pharmacologists observed that only a small number of mutations would cause genetic diseases as most don't have any impact on the health and normal growth of organisms. As molecular geneticists observed that mutations have the potential to alter the DNA sequence of gene but it fails to alter the function of protein regulated by that gene. Furthermore, they also noted that the gene mutations responsible for genetic disease are repaired by certain enzymes prior to gene expression and production of the altered protein. Contemporary findings on the mutation of DNA damages revealed that the biochemical pathways of the mutational changes are tracked down by the enzymes to repair the damages and to prevent mass level alteration/mutation of genes. Sometimes, a small mutational change leads to a new form of proteins that help to protect the extant and the derived population in future to protect it from the chance of being infected by pathogenic or disease causing bacteria or protozoa etc. As the somatic mutations alter the non-reproductive cells, it is never transferred from one generation to another e.g. the golden half of the red coloured apple, caused by somatic mutation and which is not carried over by the offsprings (Griffiths et al, 2000). Whereas, germline mutations transmitted from the parent to the embryo is found to have greater effect as witnessed in phenotypic changes in the offspring population and it can be further studied in the following categories (Griffiths et al, 2000):

a. No change occur in the phenotypes: In this category, little or no
    phenotypic change is observed in the offsprings in post-mutational
    stages. Such mutational changes might happen in two ways—either
    the mutational changes in the DNA might be in the non-coding
    region or there might be a mutational change in protein coding
    region that does not lead to any amino acid substitution, deletion or
    addition that could alter the sequence of amino acids as well as
    structure and function of the protein.

b. A small change occur in the phenotypes: A single mutational
    change causes the ears of a cat to curl backward. e.g. the "floffed-
    over ears" of the Scottish Fold breed and the "reverse curled ears"
    of the American Curl breed is caused by the mutation of the genes
    in the cartilage tissues responsible for shaping the ears and forming
    a cushion pad at the end of bones (Kelley et al, 2015).

c. A fairly large change occur in the phenotypes: The mutational
    changes like DDT resistance in insects is referred to as the effect of
    single mutation with huge impact. In some cases, the single
    mutational effect could even be lethal to kill the organism.

Molecular geneticists observed that the DNA molecule constitutes
different type of genes, of which some genes have a less powerful impact on
mutation. However, a stretch of DNA could control (those are recognized as
control genes) and switch on the function of other genes. So the indirect
mutational impact on control genes is more intense than the impact of direct
mutation on less important genes mutated as a result of alteration in the
DNA sequence. The mutation of control genes could lead to alteration of
vast array in structural and functional changes in proteins (like chain
reactions), thereby, having a more intense, enigmatic role in evolution.

Let's review the biological role of mutation so as to determine whether
it acts as the cradle of origin and evolution of biological organisms. In 1890,
Hugo de Vries, the Dutch botanist strongly advocated in favor of
"Saltationist theory of evolution by mutation" which openly criticized the
"Darwinian doctrine of natural selection" and denounced it as the ultimate
principle of biological evolution (deVries, 1900-03). The ideas of deVries
got an extended lifeline till the Neo-Darwinian principle evolved during
1930-1940s which openly promoted the role of mutation as the key impetus

of natural variations that natural selection worked on for the evolution of a diverse array of life on Earth. The contemporary work of the scientists on mutational experiments revealed that it is the physical manifestation of the "Universal mechanical damage" that alters the molecular footprints of life (Williams et al, 2008). The effect of mutation is found to be deleterious and responsible for ageing, cancer, disease etc. of multicellular organisms on Earth as it causes "inexorable genome decay" (Eccleston and Dhand, 2006; Niedernhofer et al, 2006; Kudlow et al, 2007; Williams, 2008). Going one step forward the molecular biologists who criticized the early perception of mutation's role in "Life's Diversity," expressed their concern with little satire that "beneficial side effects of the damaging mutations" could be cited with examples such as CCR5-delta32 beneficial mutation (Lamb, 2006) where deletion of 32 base pairs in a human chromosome offers resistance to HIV infection in homozygous state and delaying of setting of HIV infection in heterozygous state (Guilherme and Pacheco, 2007); and the suppression of the pelvic spine armour of the stickleback (Carroll, 2008). Evolutionary biologists hypothesized that in comparison to natural selection, which is a very slow-paced, long-lasting sustainable biological process of evolution, the ultimate outcome of selective fitness of natural selection is hardly witnessed within a single human lifetime. Whereas, mutation is found to be much faster bio-chemical mechanism, the effects of which have prominently been witnessed in a single human lifespan. The mode of action of mutation is so fast that a slow evolutionary process—natural selection—does not have much power to wipeout the minutely dispersed and ramified outcomes of mutations and eliminate those instantaneously. The evolutionary biologists shared the idea of reproductive biologists who were of the opinion that our reproductive cells are prone to damage just like our body cells (as it does not have any better immune potential than body cell) and that an extent of mutation between 10,000 to few million would be enough to push the entire human race to extinction in next hundreds to thousands years in this era of Anthropocene (Williams, 2008).

Though Neo-Darwinian theory has defined mutation as a unique biological event that is supposed to act as the prime thrust to enact biological variation, contemporary studies defined it as a purely physical event, a cumulative dynamic of solid and liquid interactions at a cellular level. The liquid part of the cell is an aqueous solution (mostly water), whereas, the solid biomolecular particles comprises biopolymers of carbohydrates, fats, proteins and nucleic acids (DNA, RNA) with diverse size, shape, structural orientation and functions. So, according to the Newton's law of motion, if there is not enough force to interfere, the non-moving molecule remains at a stand-still and the moving particle will remain moving forever and the movements of the biomolecules are assisted by ATP or GTP bio-energy fuels most of the time. Eventually, the interfering force, which was found to be of two types either shearing or ploughing force, is a witness in intercellular or intracellular logistics of biomolecules (Legett et al, 2005). The scientists noticed that most of the long-chained or multi-threaded biomolecular molecules are tightly integrated with hydrogen bonds (Williams, 2008). As a result of the interaction between frictional force and internal collision among biomolecules, the structural integrity of the biomolecules is altered e.g. DNA molecule. Scientists observed that replication and transcription related alteration of DNA molecule like movement from nucleus to cytoplasm, unwinding of the double helix, separation of double helix as template strands, complementation of bases to form complete strands of DNA and RNA molecules, and re-zipping of the newly-built DNA threads to the template strands, rewinding etc. involved some structural damage and no damage repair process was found to be 100% perfect which led to more chances of improperly repaired or evolved DNA in the mutated form (Lehninger, 1987; Griffiths et al, 2000). Conversely, the mutated or altered DNA, as well as improperly transcribed RNA are translated to mutated proteins in turn.

Dr. John C. Sanford, the famous molecular biologist, adjunct professor at Duke University, USA, redefined the Neo-Darwinian theory stating that natural selection acted as biological saviour of evolutionary process, screening out the deleterious mutation and promote the favourable ones. Professor Sanford used an analogy where natural selection mechanism is a sieve with big pores through which a number of mutations go through without being detected or filtered out by the process of natural selection. According to contemporary observations of the molecular biologists, if the mutation rate per person, per generation of any population is around 1, then the entire population supposed to be mutant and natural selection would fail to degenerate the entire population (Williams, 2008). Later, the population biologists observed that the deleterious mutation rate in Drosophila is around 1.2 around 3 in the human species (Williams, 2008). Though rate of mutation varies in the eukaryotes, scientists empirically ascertained that "multicellular genomes" have gone through the "inexorable genome decay" from mutational damage that natural selection could not repair (Baer et al, 2007; Eyre.-Walker and Keightley, 2007).

August Weissman (1893) in his work "The Germ-Plasm: A theory of heredity" clearly distinguished between body cells (soma) and the reproductive cells or germ-line cells (germplasm). The germ-line cells were predicted to be more resistant to mutation than the body cells. Later, Weisman's "germ-plasm hypothesis" put forth "Immortal DNA Strand Hypothesis" which states that during replicating the DNA of embryonic stem cell, the "Old DNA Strand" remained integrated to the "Self renewing mother stem cell," while the newly built strand moved down to differentiate between body cell (Cairns, 1975). The scientists defined this "Old Strand," which remained strategically error free of copy-related errors caused by mutational change, as "Immortal DNA Strand" (Williams, 2008). However, contemporary researches of Mark J. Kiel and his associates at the Center for Stem Cell Biology, University of Michigan, USA, tested the "Immortal DNA Strand Hypothesis" and observed that the stem cells producing blood didn't asymmetrically segregate the chromosomes. Rather, they segregate randomly which indirectly proves "Immortal DNA Strand Hypothesis" wrong (Williams, 2008). The molecular biologists came to the conclusion that as stem cells are not endowed with preferential advantage, the germ cells don't have preferential advantageous myth like "better mutation resistant" than body cells (Maklakov and Immler, 2016). So molecular damages, caused by mutational changes have the same effect on both somatic and germ cells of any organism at the species level. The worrisome part of the deleterious effect of the mutation which is studied through computer simulation model in the highest evolved species on Earth, the human species, predicted that around a few million mutations would push the human species to the brink of extinction and is likely to occur in "hundreds of thousands of years." This time span is assumed to be much lesser than the evolutionary time scale (Williams, 2008; Wylei and Shakhnovich, 2012; Porterfield, 2018). If deleterious effect of mutation is found to be so damaging in terms of sustainability of species diversity on Earth, where the highest evolved human species could be wiped out much faster than its normal evolutionary time-span, the fate of the other less advanced eukaryotic organisms is easy to imagine (Porterfield, 2018). Like "genetic erosion," mutation has been considered as a deleterious biological process and the metaphor "like rust eating away the steel in a bridge" is very much applicable (Williams, 2008; Bijlsma and Loeschcke, 2012).

References

Baer, C. F., Miyamoto, M. M., and Denver, D. R. (2007). Mutation rate variation in multicellular eukaryotes: causes and consequences. Nat. Rev. Genet. 8:619–631.

Bertram, J. S. (2000). The molecular biology of cancer. Mol Aspects Med. 21(6):167-223. doi: 10.1016/s0098-2997(00)00007-8.

Bijlsma, R., and Loeschcke, V. (2012). Genetic erosion impedes adaptive responses to stressful environments. *Evolutionary applications*, 5(2): 117–129.
https://doi.org/10.1111/j.1752-4571.2011.00214.x

Cairns J. (1975). Mutation selection and the natural history of cancer. Nature. 255(5505): 197-200. doi: 10.1038/255197a0.

Carroll, S. B. (2008). Evo-devo and an expanding evolutionary synthesis: a genetic theory of morphological evolution. Cell 134:25–36. (10.1016/j.cell.2008.06.030)

Carlice-dos-Reis, T., Viana, J., Moreira, F., Cardoso, G., Guerreiro, J., Santos, S., Ribeiro dos Santos, A. (2017). Investigation o mutations in the HBB gene using the 1.000 genomes database. PLOS One 12. E0174637. 10.1371/journal.pone.0174637.

De Vries, H. (1901-1903). Die mutationstheorie. Versuche und beobachtungen über die entstehung von arten im pflanzenreich. Leipzig & Viet Comp.

Eccleston, A. and Dhand, R., (2006). Signaling in cancer. *Nature* 441(7092):423.

Eyre.-Walker, A. and Keightley, P.D. (2007). The distribution of fitness effects of new mutations. Nature Reviews Genetics 8 (8): 610-618.

Fiethen, L., and Starlinger, P. (1970). Mutations in the galactose operator. Mol Gen Genet 108:322–330.

Griffiths, A.J.F., Miller, J.H., Suzuki, D.T., Lewontin, R.C. and Gelbart., W. M. (2000). An Introduction to Genetic Analysis. 7th edition. New York: W. H. Freeman. Available from:
https://www.ncbi.nlm. nih.gov/books/ NBK21766/

Guilherme, A. and Pacheco, F., (11th August 2007). CCR5 receptor gene and HIV infection,
<www.cdc.gov/genomics/hugenet/factsheets/FS_CCR5.htm>

Jacob, F., and Monod, J. (1962). On the regulation of gene activity. *Cold Spring Harbor Symposia on Quantitative Biology* 26: 193–211.

Jones, P. L., Veenstra, G. J., Wade, P. A., Vermaak, D., Kass, S. U., Landsberger, N. Strouboulis, J. and Wolffe, A.P. (1998). Methylated DNA and MeCP2 recruit histone deacetylase to repress transcription *Nat Genet* 19: 187– 91 doi:10.1038/561.

Kelly, L. A., Loza, A., Lin, X., Schroeder, E. T., Hughes, A., Kirk, A., and Knowles, A. M. (2015). The effect of a home-based strength training program on type 2 diabetes risk in obese Latino boys. J Pediatr Endocrinol Metab. 28(3-4):315-22. doi: 10.1515/jpem-2014-0470.

Kudlow, B. A., Kennedy, B. K. and Monnat, R. J. Jr, (2007). Werner and Hutchinson–Gilford progeria syndromes: mechanistic basis of human progeroid diseases. *Nature Reviews Molecular Cell Biology* 8:394–404.

Lamb, A., (2006). C C R 5–delta32: a very beneficial mutation. *Journal of Creation* 20(1):15.

Lehninger, A. L. (1987). Principles of Biochemistry. CBS Publishers and Distriutors, New Delhi.

Lepage, P., and de Perre P, V. (2012). The immune system of breast milk: antimicrobial and anti-inflammatory properties. Adv Exp Med Biol. 743:121-37.

Leroi, A. M. (2003). Mutants: On Genetic Variety and the Human Body. Viking, New York. Pp 448.

Lodish, H., Berk, A., Zipursky, S. L., Matsudaira, P., Baltimore, D. and Darnel, J. (2000). Molecular Cell Biology. 4th edition. New York: W. H. Freeman. Available from:
https://www.ncbi.nlm.nih.gov/books/NBK21475/

Maklakov, A. A. and Immler, S. (2016). The Expensive Germline and the Evolution of Ageing. Curr. Biol 26(13): PR577-R586.
https://doi.org/10.1016/j.cub.2016.04.012

Niedernhofer, L.J., Garinis, G.A., Raams, A., Lalai, A.S., Robinson, A.R., Appeldoorn, E., Odijk, H., Oostendorp, R., Ahmad, A., van Leeuwen, W., Theil, A.F,, Vermeulen, W., van der Horst, G.T,, Meinecke, P., Kleijer, W.J., Vijg, J., Jaspers, N.G., and Hoeijmakers, J.H. (2006). A new progeroid syndrome reveals that genotoxic stress suppresses the somatotroph axis. Nature. 444(7122):1038-43. doi: 10.1038/nature05456. PMID: 17183314.

Pandey, S.N. (2017). Principles of genetics. Ane Books.

Paulson, K.G., Voillet, V., McAfee, M.S., Hunter, D.S., Wagener, F.D., Perdicchio, M., Valente, W.J., Koelle, S. J., Church, C.D., Vandeven, N., Thomas, H., Colunga, A.G., Iyer, G.J., Yee, C., Kulikauskas, R., Koelle, D. M., Pierce, R.H., Bielas, J.H., Greenberg, P.D., Bhatia, S., Gottardo, R., Nghiem, P. and A. G. Chapuis, A.G. (2018). Acquired cancer resistance to combination immunotherapy from transcriptional loss of class I HLA. *Nat Commun* 9: 3868.
https://doi.org/10.1038/s41467-018- 06300-3

Porterfield, A. (30th November, 2018). Andrew Porterfield Discusses CRISPR in the Grocery Store.
https://www.synthego.com/blog/andrew-porterfield- podcast

Sawyer, S. A., Parsch, J., Zhang, Z., and Hartl, D. L. (2007). Prevalence of positive selection among nearly neutral amino acid replacements in Drosophila. Proc Natl Acad Sci U S A. 104(16): 6504-10. doi: 10.1073/pnas.0701572104.

Shaw, K. (2008) Negative transcription regulation in prokaryotes. *Nature Education* 1(1):122.

Silver, H. K., Kiyasu, W., George, J., and Deamer, W. C. (1953). Syndrome of congenital hemihypertrophy, shortness of stature, and elevated urinary gonadotropins. Pediatrics. 12(4):368-76.

Stadtman, E. R., Cohen, G.N., LeBras, G., and de Robichon-Szulmajster, H. (1961). Feed-back Inhibition and Repression of Aspartokinase Activity in *Escherichia coli* and *Saccharomyces cerevisiae*. J. Biol. Chem. 236:2033-2038.

Stryer, L. (1988). Biochemistry. W.H. Freeman & Co. Ltd. 1089p.

Weismann, A. (1893). Das Keimplasma: eine Theorie der Vererbung, [1892]. Translated by W. Newton Parker and Harriet Rönnfeldt as The Germ-Plasm: a Theory of Heredity. New York: Scribner. http://dx.doi.org/10. 5962 /bhl.title.25196 (Accessed March 7, 2014).

Weitzel, J.N., Lagos, V.I., Herzog, J.S., Judkins, T., Hendrickson, B., Ho, J.S., Ricker, C.N., Lowstuter, K.J., Blazer, K.R., Tomlinson, G., and Scholl, T. (2007). Evidence for common ancestral origin of a recurring BRCA1 genomic rearrangement identified in high-risk Hispanic families. Cancer Epidemiol Biomarkers Prev. 16:1615–1620.

Williams, A. (2008). Mutations: evolution's engine becomes evolution's end! Journal of creation 22(2): 60-66. https://creation.com/article/5884/

Williams, C.A., Dagli, A. and Battaglia, A. (2008). Genetic disorders associated with macrocephaly. Am J Med Genet Part A 146A:2023–2037.

Wylie, C. S., and Shakhnovich, E. I. (2012). Mutation induced extinction in finite populations: lethal mutagenesis and lethal isolation. PLoS Computational Biology 8(8): e1002609.

Chapter 5

Reviewing Biochemical Basis of Inheritance, Concept of Gene, Origin, and Evolution of Cell Organelle

Genes are subsets of the data set defined by the nucleotide sequence of DNA. To qualify as a gene, the data subset must be so formatted that it can be interpreted by an organism into a distinct biochemical activity. An important implication of this definition is that, because biochemical activities are distinct and chemically separable from other such activities, genes may become manifest as distinct and distinguishable, biological phenotypes.

---- John Hewitt

Around 1590, Dutch spectacle makers, Hans Janssen and his son Zacharias Jansen, experimented with multiple lenses in a long tube to observe the smallest objects which was neither visible to the naked eye nor under magnifying glasses and thus started the history of microscopy in the world of science. However, Dutch inventor Anthony Leeuwenhoek is recognized as the father of microscopy. He initiated scientific observation of yeast, bacteria, red blood cells and different types of invisible organisms (to the naked eye) under the microscope which led to a glorified regime of biologists who discovered the microscopic structures of cells. In 1885, the substantial evidence of chromosomes, the key component of nucleus, was found to be hereditary material as well as genetic material which was discovered by the scientific communities under microscopic observation. Development of microscopic techniques along with advanced histochemical techniques helped to discover the biochemical constituents of the chromosome, 50% of which was constituted protein and the remaining 50% was found to be made up of DNA (by weight). The discovery of the double helix model of DNA by Watson and Crick (1953a) not only elucidated the demonstration of semi-conservative mode of replication of DNA but also revealed that DNA carries the genetic information of the concerned organism used at the time of reproduction of cell. Initially, it was understood that the entire hereditary material was contained within DNA of the cell nucleus but next level of sub-microscopic studies with better histochemical techniques helped to decipher that extra-hereditary information are also present in the minute content of two cellular organelles: mitochondria, which is present exclusively in animal cell and mitochondria along with chloroplast, those of which are present in plant cell (Alberts et al, 2002). Regarding the evolutionary origin of DNA in cellular organelles like mitochondria and chloroplast, the endosymbiotic hypothesis helped to understand that those organelles supposedly originated as free-living prokaryotic cells like bacteria or archaea and were later to be found living in larger host cells belonging to different group of organisms (phylogenetically related under "tree of life") (Timmis et al, 2004). At one point of time they evolved together and could not structurally or functionally exist independently as a result of symbiosis as well as inclusion of those organelles in higher group of host organisms, thus, the eukaryotes evolved out of the prokaryotes (Timmis et al, 2004). The lack of fossil evidence is the main impediment to establish evolutionary relationship between prokaryotes and eukaryotes which prompted the scientists to look for molecular evidence to establish a relationship between them. There are a number of hypothesis regarding origin and evolution of prokaryotes and eukaryotes, however, at a cyto-genetical distinction between these two groups of cell types are well defined. The smaller-sized (1-5 micrometers), unicellular, prokaryotic cells are characterized by circular DNA (in free floating condition in cytoplasm) having no nucleus and membrane-bound organelles (e.g. bacteria, archaea etc.), whereas, the larger (10-100 micrometers) eukaryotes are found to be unicellular or multicellular structures characterized by possession of linear DNA (found in nucleus region) having distinct nucleus and membrane-bound organelles like chloroplast, mitochondria, golgi bodies and endoplasmic reticulum etc. Digging dipper into the endosymbiotic origin and evolution of cell organelles, it was construed that the unicellular, amoeboid eukaryotes did not evolve with photosynthesizing potential, instead, it engulfed bacteria having such potential and such evolution of mutual symbiotic association led to the formation of chloroplasts (Sharma, 1985, Timmis et al, 2004). The ultrastructural resemblance of lamellae of chloroplast to the photosynthetic bacteria render the evolutionary biologists to stand behind the endosymbiotic hypothesis (Cavalier-Smith, 2000). An identical extrapolation like symbiotic evolution of amoeboid form of cellular entity with non-photosynthesizing bacterium led to the formation of mitochondria (Saccone et al, 2000; Cavalier-Smith, 2006). Molecular geneticists ascertained that the protein content of mitochondrial region and the mitochondrial envelope are controlled by the nuclear genes of the concerned cell, rather than the wide array of proteins in mitochondria which are not controlled by the genes (specifically 30 genes) mapped in the mitochondrial DNA (Taanman, 1999; Saccone et al, 2000). On the other hand, scientists involved in mapping 300 genes of chloroplast DNA, observed that those genes of chloroplast DNA have more control over nuclear genes in synthesizing the chloroplast protein (Daniell et al, 2016). Later, a group of scientists came up with a new hypothesis which slightly modified the earlier endosymbiotic hypothesis and patronized the origin of eukaryotes from the non-photosynthesizing prokaryotes to respiratory genes but it did not have any substantial structural and the biochemical evidence proved to be a valid theory (Lopez-Garcia and Moreira, 2015). At last the evolutionary biologist came up with a composite hypothesis and a scientific compromise was made. This hypothesis suggested that instead of evolving from one another, both distinctive cellular forms evolved out of a common progenitor and supposedly one took a lead in course of time. Though, finding out of the exact time of origin and evolution of the prokaryotes and eukaryotes was a difficult phase to track in the evolutionary history of cellular entity on Earth, scientists were able to figure out that the prokaryotes with simpler genetical organization (like long chains of DNA molecule) evolved around 3 billion years ago, whereas, the highly diversified eukaryotes with complex cellular and genetical organization, may have evolved not more than one billion year ago (American Chemical Society National Historic Chemical Landmarks., November, 2009; Lopez-Garcia and Moreira, 2015). The most common feature between these two groups of cellular entity is that the universal genetic code was found to be functional between the two distinct entities which helped molecular and evolutionary biologists to focus their studies on the biochemical basis as well as hereditary unit of life, the gene.

In comparison to the nuclear organization of prokaryotes, if we review the nuclear structural and functional pattern of eukaryotes, we would notice that the nuclear membrane in eukaryotes act as a shield to protect chromosomes from direct environmental interference and abrupt changes (chance of mutagenic susceptibility etc.) by segregating it from the cytoplasm (Merchant and Favor, 2015). Such protective role of the membrane has indirectly limited the logistics of gene functioning (ultimate coding potential of the language of nucleic acid to translate protein synthesis) as well as gene enzyme action or cistron-polypeptide action as the movements of important molecular instruments such as rRNA, mRNA are restricted through the porous nuclear membrane (Cooper, 2000). Geneticists noticed that the membrane-bound organelles in eukaryotes like chloroplast and mitochondria possessing round-shaped, self-replicating, double-stranded, DNA structures, normally found in the prokaryotic cells of bacterial and archaea were known as plasmids (Sharma, 1985; Merchant and Favor, 2015). Furthermore, the investigation of ribosomal complex of chloroplasts and mitochondria in eukaryotic cells was found to be identical to the 70S ribosomal entity of prokaryotes rather than 80S unit of eukaryotes, thus, helping evolutionary biologists to stand behind the hypothesis of endosymbiotic origin of eukaryotes out of the mother stock of prokaryotes in the course of biological evolution of life on Earth (Cooper, 2000).

It was observed that DNA replication in the nucleus region followed by DNA dependent mRNA synthesis or transcription has paved the way to decode the language of nucleotide to protein. The mRNA normally travels from the nucleus region and through the cytoplasm to be integrated with ribosome and carry out protein translation. In the process of protein translation, the nucleotide sequence decodes three consecutive mRNA at a time and translates it into a specific amino acid. So, RNA plays a cardinal role in the mediation of sequencing nucleotides and translating it to amino acids to synthesis a specific protein molecule. Technically, a particular location of DNA sequence specify a particular protein which is called gene. The meticulous studies on the DNA of the eukaryotic chromosome was structurally constituted with operator and repressor genes and the action of DNA-dependent polymerase enzyme and production of mRNA triggered a specific gene, known as operon (Sadava et al, 2011). Careful observation by molecular biologists ascertained that at the time of production of mRNA from the DNA template, it was found to be equal to the length of the parent genes, complementary to the DNA nucleoid strand, (template) but its translocation from nucleus to ribosome—located in cytoplasm—made it shorter. Such shortening of the length of the mRNA of the primary transcript was genetically recognized as "processing of messenger" (Sadava et al, 2011). However, at the end of the transcription or mRNA formation, a long chain of Adenine bases, called polyadenylation, was attached at the end of the mRNA to control its logistics which was to be integrated to the ribosomes. Whereas, the methylated sequence at the end of mRNA prevent it from forming any extra attachments with more than one tail, recognized as capping (Sadava et al, 2011). As the ultimate shorter segment of mRNA, which was found to be shorter than the primary transcript, it supposedly possessed lesser genes than the parental form. It coordinated coding as well as protein translation where a number of amino acids were supposed to be missing in the polymerization of protein molecules. Hence, a hypothetical chance of mutation existed where parental enzymatic configuration or the representation and expression of proteins determined the phenotypic expression of contemporary generations (the offsprings). In reality, no such altered structural or functional forms of protein have been found in the translation process determining that splitting of segments of mRNA, which might lead to the loss of some genes, did not alter the coding pattern. It also showed that genes contained by the split segment might not carry any essential code (which should better be recognized as non-coding region of DNA) that could affect the polymerization of proteins as well as alter the sequencing of amino acids in the biosynthesis of proteins (Baudrimont et al, 2017).

Hence, according to the interrupted-gene theory or split-gene theory, in the DNA molecule, the interspersing sequence of nucleotides constituting non-coding genes called introns was to be found in between the segments of nucleotides constitute functional genes which are recognized as exons. The coding regions of the DNA nucleotides, containing exons, executed the expression of proteins via mRNA, whereas, the functional role of the interspersing, non-coding regions of the DNA, which constituted introns were found to be non-functional. This split-gene hypothesis was jointly promulgated by Richard J. Roberts and Phillip A. Sharp in 1977; eventually, this miracle discovery would help the scientist duo to achieve Nobel award in Physiology and Medicine in 1993 (Altman, 1993). The non-coding DNA sequence of the entire molecule, which constituted introns are known as non-coding DNA or ncDNA or junk DNA. It was further observed by the scientists that non-coding DNA segments are involved in splicing introns-forming precursor of mRNA, which is required for transcription of the entire

RNA. It needs to be mentioned here that introns are spliced out by spliceosome which carry in-built molecular scissor, or excision enzyme, called restriction endonuclease. This makes the mRNA shorter than the original length of primary transcript (Pray, 2008a; Jo and Choi, 2015; Lepennetier and Catania, 2016). The segmented coding regions of the mRNA are joined or glued together by an enzyme called polynucleotide ligase (Shuman and Lima, 2004). The shorter mRNA integrates with the ribosome to act as a mRNA strand responsible for protein translation and polyadenylation or tailing short mRNA strand by helping to sync with ribosome in the process of protein translation (Lutz and Moreira, 2011). It should be mentioned here that 98% of human genes are interrupted or non-coding genes, hence, only 2% of the human genome constituted functional coding regions responsible for protein translation. It has further been ascertained that 50% of the hereditary disease in human beings is due to erroneous splicing of introns in interrupted genes (Ward and Cooper, 2010; Prabakaran et al, 2014).

The absence of primary transcript, shortened transcript, split genes and introns in the prokaryotic organisms questioned the validity of the hypothesis of prokaryotic origin, instead, the parallel evolution of prokaryotes and eukaryotes from the common mother-stock was put forward (Gu, X, 1997). Eventually, molecular biologists came up with a contemporary concept of gene, particularly, including the property of splicing of introns, interrupted gene and split genes to reconcile endosymbiotic origin with evolution of membranous cell organelles, like chloroplast (one with photosynthesizing potential) and mitochondria (having respiratory and energy yielding potential) (Sharma, 1985). The origin, evolution, and diversification of eukaryotes is apparently a result of reticulate evolution which leaped huge amount of time and space and created an advanced group of organisms (structurally and functionally) with diverse array of proteins. It is understandable that diverse array of protein synthesis as well as advance level of protein translation must be encoded by intensive functional mode by DNA polynucleotides, mediated via more precise functional mode of mRNA.

The universal identity of the molecular consistency of biological organism is recognized by the role of nucleic acid and the biochemical basis of inheritance. While the consistency of its genetic profile is ascertained by the dominance of sequence of DNA content at a cellular level which plays a cardinal role in self-replication on its course of differentiation in the same generation and reproduction of the following generations (Bailey et al, 2015; Bailey, 2019). When streamlining flawless DNA replication then next level of transcription of mRNA from DNA and the process of protein translation, coded by the DNA molecule via mRNA molecule is done through sequencing of amino acids. This is responsible for synthesis of protein (enzyme specificity) and factors like alteration of population size and mutational variation to end up with changes in phenotypic expression and are found to be responsible for repeated evolution (Bailey et al, 2015). Though double helix model of Watson and Crick (1953a) was found to be one of the best model to depict the functional and structural entity of DNA and gene, the beginning of the 20th century saw critical review by the molecular geneticists for following reasons:

a. Speyer (1965) studied the DNA replication experiments of *E.coli* in cell-free cultures and observed that application of the altered (mutated) DNA polymerase yielded the mutated form of DNA profile in the new generation of *E.coli* population, rather than replicating the archetype DNA helix of the ancestral type. Most surprisingly, the scientists noticed that the mutant DNA polymerase were involved in controlling the G : C replication rather than A : T replication (Speyer, 1965). So, it was deduced that the replication of parental DNA molecule would be successful when the consistency of the archetype DNA template could interact with unaltered (non-mutated) DNA polymerase enzyme.

b. The molecular biologists critically reviewed that DNA guided protein translation was found to be a complex process as it required assistance of ribosomes, mRNAs and tRNAs to which attachment of amino acids were required and fulfilled by the synthetase or ligase enzymes, involved in binding two substrates and synthetase using high energy nucleosides (like AATP, GTP etc.) to procure energy. Leder and Nirenberg (1964) in their experiment to decipher the tri-fold nature of genetic code carried out ribosome binding assay in which they successfully passed various combinations of mRNA through a filter of ribosomes. They observed that triplet codons render the binding for specific tRNAs to the ribosomes (American Chemical Society National Historic Chemical Landmarks, 12th

November, 2009). The scientists coordinated this experiment as they intended to associate the tRNAs to a specific amino acid as they tried to determine the triplet mRNA sequence or sequence of codons which coded for each amino acid. The uniqueness of this experiment on applied aspects showed that a specific codon had different binding capacities tried against amino acid-tRNA complexes of different species like *E. coli*, *Xenopus* liver or guinea-pig liver (Wang and Schulz, 2005). So sequencing of amino acids, as well as derivation of distinct protein molecules was found to be dependent upon the interactions of specific mRNA and amino acid-tRNA complex catalyzed by tRNA amino acid synthetase.

c.   Molecular biologists further observed that the catalyzing potential of protein synthetase helped in successful execution of distinct pattern of sequencing of amino acids to form protein polypeptide chain. It was observed by the scientists that the synthetase, yielded by different species, having different extent of potential to sequence the same amino acids to synthesize protein polypeptide. From contemporary molecular biological point of views, it is construed that biochemical specificity of protein depends upon the specificity of synthetase, which synthesizes it (Weimar et al, 2002). The recent advancement of the molecular geneticists and the genomicists in the domain of protein biosynthesis has expand the erstwhile horizon of genetic code promulgated in the 90s by breaking the limitations of the biosynthesis of proteins, encoded by the permutation and combination of 20 amino acids and stepped out to look for the unique, under-explored and unexplored potential of proteins in different organisms. It has been substantiated furthermore by the contemporary findings of the molecular biologists as they have observed, *"For most proteins, modifications are largely restricted to substitutions among the common 20 amino acids. Herein we describe recent advances that make it possible to add new building blocks to the genetic codes of both prokaryotic and eukaryotic organisms. Over 30 novel amino acids have been genetically encoded in response to unique triplet and quadruplet codons including fluorescent, photoreactive, and redox-active amino acids, glycosylated amino acids, and amino acids with keto, azido, acetylenic, and heavy-atom containing side chains.*" (Wang and Schulz, 2005).

d. It was also observed by the contemporary molecular biologists that protein specificity, related to ribosomal integration, as well as polysomic model of protein translation was not a universal model. As scientists worked on the mutational studies of genome of *E.coli,* they observed that two alternative mutations lead to the formation of two different strains of *E.coli* having two distinct type of ribosomes. Thus, two distinct types of mutant strain ultimately lead to two distinct types of phenotypic expression resulting in two distinct protein translation mechanisms of the same codon (Alberts et al, 2002).

By referring to the weakness in Watson and Crick's (1953a) structural and functional Double helix model, Commoner (1968) observed unidirectional mode of operation of DNA synthesis and protein synthesis. It apparently looked similar but expression of biological specificity in biological phenotype of any organism is not necessarily universal. It would stereotypically follow the path of Central dogma in molecular biology, DNA replication, DNA to mRNA transcription and protein translation, rather than biochemical specificity witnessed in the specific sequence of amino acids in the synthesized protein which partly came from pre-existing DNA and partly inherited from pre-existing proteins. However, Commoner's (1968) view was further supported by the multi-molecular hypothesis in gene therapy of Robert L. Sinsheimer's (1970) "Genetic Engineering: The modification of man," where a valid concern was expressed that suggested biochemical alternation of DNA molecule might not yield the best result in gene therapy to cure any genetically transmitted disease as manipulation of the mode of replication, transcription and protein translation would not solely determine biochemical specificity of DNA nucleotide and biochemical specificity of protein. Rather, biochemical specificity was ultimately integrated to different functional and structural constituents of DNA double helix, its mutational susceptibilities and its responses to the enzymatic actions of synthetase etc. (Weimar et al, 2002)

In the history of molecular biology, there is a new terminology that has been recently added—"Post-genomic era," which is related to the most contemporary advancement in the field of investigations of genome sequence. It consisted of development of computer-aided software/programs to analyze and retrieve the bulk of genome sequence data, sharing of the unique findings among like–minded research institution and academics and availability of the genome sequence data of a number of species with the scientific communities (Lengauer, 2000; Baxevanis and Ouellette, 2001). So, till 1995, if the advancement of molecular genetics were more individualistic in its efforts, then certain patronized research and academic institutions and availability of the fragmentary information of the sequenced data of genomes could have hardly been shared across the globe and the concept of gene would not have been known to the scientific communities. The pro-conservative advancement or the early phase of genomic research till 1995 (or 1999) was popularly recognized as "Pre-genomics era" (Lengauer, 2000). Whereas, the huge shift in the advancements in the field of genomics, like collective approaches of investigations, availability of complete form of sequenced data of genomes across the academic and research institutions, collaborative studies and prioritizing the collective reviews, made it a "pro-liberal effort" in the last ten years.

The advancement of molecular technology has changed the perception of genes from time to time. As par classical perception, gene is an integral part of the entire genome. The mutational change in it could be reflected in the altered expression of phenotype of the concerned organism. However, any stereotypic perception could not persist for an indefinite period of time and in the "post-genomic era," thus, earlier perception of gene needs to be considered with certain modification. The main conflict in adjudging the structural and functional properties of gene in accordance with the "Central dogma in molecular biology" of Crick (1957,1970) and "Split-gene model" of Richard J. Roberts and Phillip A. Sharp starts whenever the Central Dogma talks about a chain reaction where the genes stationed in DNA molecule are replicated once, before encoding RNA transcripts, followed by encoding RNA transcripts which successfully translates it to polymerization of amino acids and finally finished with formation of proteins. While, the 'Split-gene model' promotes the idea of the formation of protein by utilizing a smaller section of genome (exclusively exons), the remaining vast segment of genome (interspersing introns), popularly recognized as junk DNA, or ncDNA abstained from taking part in protein translation (Wells, 2011). The functional mode of this ncDNA still remains enigmatic to the molecular biologists. One particular group construed that ncDNA is a chunk of junk

DNA and has no significant function, whereas, another group of molecular biologists theorized that it very much engaged in diverse action of transcription to form different forms of RNA (Wells, 2011). So, split gene has radically promoted the idea of DNA to RNA transcription, resembling Central dogma but it does not promote the idea of translation of entire primary transcript of RNA (which is exact complementary entity of DNA); rather split-gene model promoted the idea of protein translation, carried out by a tiny, functionally active (coding) region of genome (Pray, 2008b). On the other hand, contemporary findings by molecular biologists show that large sequence of "non-protein-coding regions" are transcribed to RNA and they consider it as either "non-protein-coding sequence" or a transcriptional mistake (Dinger et al, 2008). Recent molecular investigation on fruit flies (*Drosophila melanogaster*) revealed that 29% of the sequence of genome constitute alternative exons of known genes, whereas, 71% of the sequence is either made up of "unknown protein coding genes" or by a number of "untranslated RNAs" (Zhong et al, 2014; Piergentili, 2020).

The careful review of the consecutive steps of "Central Dogma in molecular biology" by Crick (1958), in light of the contemporary findings of the molecular foot-print of gene, clearly elucidated that the first two steps of DNA to DNA replication and DNA to mRNA transcription followed the principle of complementary bases but the next stage of protein translation, which was stereotypically defined as translation of DNA codes via RNA to the sequential formation of amino acids or biosynthesis of proteins, might not be indispensable criteria to define gene (Piergenili, 2020). The contemporary findings in the molecular biological investigations has witnessed that the genes of tRNAs, rRNAs and the non-translated RNAs defined in the split-gene model have not gone through the process of translation. So, according to the conventional function of genes, they are supposed to code mRNAs and coordinate protein translation and are recognized as "structural genes," which was jointly promulgated as "Lactose Operon Model" in 1961 by two eminent French molecular geneticists, François Jacob and Jacques Monod from Pasteur Institute of Paris, France. In their unique "Lac-Operon" model (Jacob and Monod, 1961), there was an earnest effort to distinguish between structural gene and regulatory gene. It also further elucidated that the gene which is encoding the lactose repressor protein is called 'Regulatory gene' and defining the mode of action of this repressor gene. Jacob and Monod (1961) further stated that the regulatory gene did not work superficially like structural genes, rather, these regulatory genes indirectly control the functional pattern of the structural genes. The distinction between structural and regulatory gene has remained a grey area for molecular genetics for a long time. If we dig deeper into genetically significant DNA sequence, it leads the molecular investigation to locate the regions of the DNA that are not transcribed to protein regulators. Moreover, it was observed that it would not distinguish two forms of RNA, the first— those that are translated into proteins (e.g. enzymes, regulator proteins etc.) and the second— that are transcribed into RNAs (e.g. rRNA, tRNA and non-translated RNA etc.) but are never translated into protein. So, after traversing a long way down the trail of refinement of hypotheses, ideas and evidence to define genes in contemporary perception in molecular genetics, the erstwhile definition of genes need to be reviewed. Though the evolutionary journey of gene was initiated by Gregor Johann Mendel in 1885, the father of genetics with his "One-gene one-phenotype hypothesis," the history of genetics has gone through several shifts in paradigm, such as "One-gene one-enzyme hypothesis" proposed by George Beadle and Edward Tatum (1941), "Lac-Operon Model" of Francis Jacob and Jacques Monod (1961). The molecular perception of gene has gone through radical changes and the unique molecular footprints of gene has defined its primary role of transcribing as well as encoding DNA sequences to form the primary RNA transcript. It clearly elucidated further that the protein translation role of DNA at a post-transcription level and coordinating the protein translation via mRNA is supposed to be an extracurricular features of DNA sequences or gene. After coming a long way in the evolutionary trail of molecular genetics, the perception of gene has gone through the tertiary shift of paradigm—"The split gene model" which was promulgated by Richard Roberts and Phillip Sharp in 1977 (Altman, 1993). It clearly defined that gene expression might be a phenotypic expression or it might be expressed in the form of proteins or enzymes but protein translation is not the destination of gene, rather, it should be considered as a journey exposing the non-coding sequences of DNA and discoveries of introns, primarily, restricting the first stop-over of the journey of gene to transcription and protein translation recognized as part of an extended journey. Even after a long journey of recognition of a diverse array of structural and functional entity of genes, most molecular geneticists are still divided in defining gene as "protein coding regions" (Harrow et al, 2009). However, after going through the all composite observations on the molecular footprints of DNA coding sequences, starting from Mendelian regime of hereditary factors at the beginning of the 19ᵗʰ century to the contemporary epoch of split genes, it was required to redefine the DNA coding sequences as well as the contemporary refinement of the concept of genes. Those that have gone through the spatio-temporal refinements of perception in the field of molecular biology should further be studied under four categories (Griffiths and Stotz, 2005):

a. Translatable sequences: In this category, DNA sequences are being transcribed to mRNA and translated via mRNA to the polymerization of amino acids or biosynthesis of proteins. The proteins either involves regulation of DNA or non-DNA-related activities. So, this category of DNA sequences are found to be proactive in coding protein via transcription of RNA, followed by execution of translation via RNA.

b. Transcribable sequences: In this category, DNA sequences are transcribed into RNA (like rRNAs and tRNAs) but those transcribed RNAs don't take part in polymerization of amino acids or biosynthesis of proteins. This category of DNA sequences are successful in transcription of RNA but remain unsuccessful in translation of protein in the absence of right formation of RNA in the transcription phase.

c. Binding sequences: In this category, the DNA sequences are neither transcribed to RNA nor translated to protein but act like a binding site of regulatory genes like protein repressors or homeotic genes. Though these kinds of sequences are neither transcribed nor translated into proteins they still carry out regulation of other structural genes by forming regulatory binding sites.

d. Non-binding sequences: In this category, the DNA sequences are neither transcribed to RNA, nor translated to protein and do not acting as binding sites for regulatory molecules. These types of repetitive DNA sequences are comprised of pseudogenes, tandem repeats, transposons, spacer DNAs, retroviral inserts etc. Depending upon the integration of the non-binding sequences of the DNA to either of the main DNA elements of eukaryotes or the parasitic genetic elements like retrovirus or transposons. The non-binding sequences are further categorized as:

a.   functional/non-binding & non-coding DNA sequences

b.   non-functional/non-binding & non-coding/parasitic DNA sequences

The conventional idea of epigenetics is defined as "all molecular pathways modulating the expression of a genotype into phenotype" and after a span of refinement, it has specifically been interpreted as "the study of changes of gene function that are mitotically or meiotically heritable and do not entail a change in DNA sequence." (Dupont et al, 2009) In the epigenetic changes, genes are expressed without alteration in its DNA sequence and the non-genetic factors regulate the genes of the organism differentially (Philip, 2008). It was revealed by molecular geneticists that gene expression (of heritable phenotype) is controlled by the repressor protein, which is integrated to the silencer regions of the DNA (Kolovos et al, 2012). Basically, epigenetics gave an opportunity to study the changes of heritable phenotype by stepping out of the conventional genetic basis of inheritance (Berger et al, 2009). However, if we review the traditional principles of heredity, it involves similar processes to make the progeny resemble their ancestors. So, at the molecular level, principles of heredity ensure consistent replication of DNA coding to maintain the consistency of the profile of genes transferred from generation to generation. So heredity seems to be a conservative biological principle, which is found to be an impediment to experimental trial of evolution that attempted to come out of the stereotypic prototype in order to evolve with a unique genetic profile. So, Mayr (1969) considered that mutation, recombination, isolation and selections as "four corner-stones" of biological evolution which altered the chromosomal configurations as well as structural and functional parameters of genes. Along with four major corner-stones, "inbreeding depression" is one of the cardinal issue that influenced the extent and amplitude of biological evolution. Population biologists observed that inbreeding in a diverse population of any biological organism supposedly evolved an altered gene-pool which interacted with the genetic environment in a different way (Charlesworth and Charlesworth, 1987). Thus, the changes in the environment would alter the selective value of the concerned gene in its ambiance (Sharma, 1985). Any alteration of gene, either individual or collectively, would help to evolve the unique organism, thereby, befitting the outlay of natural selection and successfully yielding unique and favorable evolutionary lines. The establishment of the new genotype among the evolved populations have gone through trial and error of natural selection and genetic drift along with gene flow do not act in isolation, rather, work together to appear as unique natural population in stable form (Andrews, 2010).

Population geneticists further observed that natural selection acts as the cardinal impetus for biological evolution of the derived population out of ancestral lineages as it promotes evolution of dominant alleles in higher frequency as it makes the newly evolved organism meet the basic requirements of natural selection (Nosil et al, 2018). Whereas, genetic drift trigger the organisms to evolve with neutral characters and non-adaptive traits (with some non-favorable traits) to survive the course of evolution (Sharma, 1985). The famous evolutionary biologist, Dr Julian Huxley, in his collaborative work with A.C Hardy and E.B. Ford, titled "Evolution as a Process," opined that as an integral process of biological evolution, natural selection has certain limitations. He considered that natural selection is a bias to the instantaneous "biological utility" of the species and seems to be "incapable of purposeful design or foresighted planning." Naturally, the result of fitness of natural selection in a given environment, wherein the particular species belong to, is always relative. Hence, the "selective value" of a gene might be altered in tandem with the changes of genetical environment of the concerned organism. Huxley (1955) further observed that genes with following alternative traits are favored by the natural selection:

a.   Those producing the heterozygotes of high viability

b.   Those producing "viable combinations of the greatest number from different genetic background."

References

Alberts, B., Johnson, A., Lewis, J., Raff, M., Roberts, K., and Walter, P. (2002). Molecular Biology of the Cell. 4th edition. New York: Garland Science; Available from:
https://www.ncbi.nlm.nih.gov/books/NBK21054/

Altman, L.K. (12th October, 1993). Surprise discovery of 'split genes' wins noble prize.
https://www.nytimes.com/1993/10/12/health/surprise-discovery- about-split-genes-wins-nobel-prize.html.

American Chemical Society National Historic Chemical Landmarks. (12th November, 2009). Deciphering the Genetic Code.
http://www.acs.org/content/acs/en/education/whatischemistry/landmarks/geneticcode .html

Andrews, C. A. (2010) Natural Selection, Genetic Drift, and Gene Flow Do Not Act in Isolation in Natural Populations. *Nature Education Knowledge* 3(10):5

Bailey, R., Priego Moreno, S., and Gambus, A. (2015). Termination of DNA replication forks: "Breaking up is hard to do." Nucleus. 6(3):187-96. doi: 10.1080/19491034.2015.1035843.

Bailey, R. (7th October, 2019). DNA Replication Steps and Process. https:// www. thoughtco.com/dna-replication-3981005

Baudrimont, A., Voegeli, S., Viloria, E. C., Stritt, F., Lenon, M., Wada, T, Jaquet, V., and Becskei, A. Multiplexed gene control reveals rapid mRNA turnover. Sci Adv. 2017 Jul 12; 3(7): e1700006. doi: 10.1126/sciadv .1700006.

Baxevanis, A.D. and Ouellette, B.F.F. (2001). Bioinformatics: A practical guide to the analysis of the genes and proteins. Wiley Interscience. P.495.

Beadle, G. W., and Tatum, E. L. (1941). Genetic Control of Biochemical Reactions in *Neurospora*. PNAS. 27 (11): 499–506. doi:10.1073/pnas.27.11.499

Cavalier-Smith, T. (2000). Membrane heredity and early chloroplast evolution. Trends Plant Sci. 5(4):174-82. doi: 10.1016/s1360-1385(00)01598-3.

Cavalier-Smith, T. (2006). Cell evolution and Earth history: stasis and revolution. *Phil. Trans. R. Soc. B*361969–1006. doi.org/10.1098/ rstb. 2006.1842

Charlesworth, D. and Charlesworth, B. (1987). Inbreeding depression and its evolutionary consequences. Annual Review of Ecology and Systematics 18(1): 237-268

Crick, F. H. C. (1957). On protein synthesis; Manuscript. Cold Spring Harbor Laboratory Archives, SB/11/5/4.
http://libgallery.cshl. edu/items/ show/ 52220.

Crick, F.H.C. (1958). On protein synthesis. *Symp. Soc. Exp. Biol.* 12: 138–1163

Crick, F.H.C. (1970). Central dogma of molecular biology. *Nature* 227: 561-563.

Commoner, B. Failure of the Watson–Crick Theory as a Chemical Explanation of Inheritance. *Nature* 220: 334 – 340 (1968).
https:// doi. org/ 10. 1038 / 220334a0

Cooper, G. M. (2000). The Cell: A Molecular Approach. 2nd edition. Sunderland (MA): Sinauer Associate. Available from:
https://www.ncbi.nlm.nih.Gov /books/NBK9839/

Daniell, H., Lin, C .S., Yu, M., and Chang, W. J. (2016). Chloroplast genomes : diversity, evolution, and applications in genetic engineering. Genome Biol. 17(1):134. doi: 10.1186/s13059-016-1004-2.

Dinger, M. E., Pang, K. C., Mercer, T. R. and Mattick, J. S. (2008). Differentiating protein-coding and noncoding RNA: challenges and ambiguities. PLoS Comput Biol. :4(11):e1000176. doi: 10.1371/journal. pcbi.1000176.

Dupont, C., Armant, D. R., and Brenner, C. A. (2009). Epigenetics: definition, mechanisms and clinical perspective. Semin. Reprod. Med. 27: 351–357.

Griffiths, P. E. and Stotz, K. (2005). Gene. In David L. Hull & Michael Ruse (eds.), _The Cambridge Companion to the Philosophy of Biology_. Cambridge University Press.
https://philpapers.org/rec/GRIG

Gu, X. (1997). The age of the common ancestor of eukaryotes and prokaryotes: statistical inferences. Mol Biol Evol. 14(8):861-6. doi: 10.1093/oxfordjournals.molbev.a025827.

Harrow, J., Nagy, A., Reymond, A. , Alioto, T., Antonarakis, S.E., and Guigo, R. (2009). Identifying protein-coding genes in genomic sequences. Genome Biol 10: 201.
https://doi.org/10.1186/gb-2009-10-1-201

Huxley, J. (1955). Morphism and evolution. Heredity. 9: 1–52. doi:10. 1038 / hdy.1955.1

Jacob, F. and Monod, J. (1961). Genetic regulatory mechanisms in the synthesis of proteins. J Mol Biol. 3 (3): 318–56. doi:10.1016/S0022-2836(61)80072- 7

Jo, B. S., and Choi, S. S. (2015). Introns: The Functional Benefits of Introns in Genomes. Genomics Inform. 13(4):112-8. doi: 10.5808/GI.2015.13.4.112.

Kolovos, P., Knoch, T.A., Grosveld, F.G., Cook, P.R., and Papantonis, A. (2012). Enhancers and silencers: an integrated and simple model for their function. *Epigenetics & Chromatin* 5: 1.
https://doi.org/10.1186/1756- 8935-5-1

Leder, P. and Nirenberg, M.W. (1964). RNA Codewords and Protein Synthesis, II. Nucleotide Sequence of a Valine RNA Codeword. PNAS. 52 (2): 420–427. doi:10.1073/pnas.52.2.420

Lengauer, T. (2000). Bioinformatics: From the Pre-genomic to the Post-genomic Era. ERICM News No. 43.

https://www.ercim.eu/publication/Ercim_News/enw43/lenggauer.html

Lepennetier, G., and Catania, F. (2016). mRNA-Associated Processes and Their Influence on Exon-Intron Structure in *Drosophila melanogaster*. G3 (Bethesda). 6(6): 1617-26. doi: 10.1534/g3.116.029231.

López-García, P., and Moreira, D. (2015). Open Questions on the Origin of Eukaryotes. Trends Ecol Evol. 30(11):697-708. doi: 10.1016/j.tree.2015. 09.005.

Lutz, C. S., and Moreira, A. (2011). Alternative mRNA polyadenylation in eukaryotes: an effective regulator of gene expression. Wiley Interdiscip Rev RNA. 2(1):23-31. doi: 10.1002/wrna.47.

Mayr, E. (1969). The biological meaning of species. Biol. Journ. Linn. Soc. 1:3. https://doi.org/10.1111/j.1095-8312.1969.tb00123.x

Merchant, R.G. and Favor, L.J. (2015). How Eukaryotic and prokaryotic cells differ. Britannica Educational Publishing. 64 p.

Nobel Media AB 2020. (18th Nov 2020): Press release. NobelPrize.org. https://www.nobelprize.org/prizes/medicine/1993/press-release/

Nosil, P., Villoutreix, R., de Carvalho, C. F., Farkas, T. E., Soria-Carrasco, V., Feder, J. L., Crespi, B.J., and Gompert, Z. (2018). Natural selection and the predictability of evolution in *Timema* stick insects. Science. 359(6377):765-770. doi: 10.1126/science.aap9125.

Phillips, T. (2008). The role of methylation in gene expression. *Nature Education* 1(1):116

Piergentili, R. (2020). Y RNA in cell cycle progression and cancer. Atlas of Genetics and Cytogenetics in Oncology and Haematology 24(10):379-386. DOI : https://doi.org/10.4267/2042/70820

Prabakaran, S., Hemberg, M., Chauhan, R., Winter, D., Tweedie-Cullenm R.Y., Dittrich, C., Hong, E., Gunawarden,, J., Steen, H., Kreiman, G., and Steen, J. A. (2014). Quantitative profiling of peptides from RNAs classified as noncoding. Nat Commun. 5:5429. doi: 10.1038/ncomms6429.

Pray, L. (2008a). What is a gene? Colinearity and transcription units. *Nature Education* 1(1):97.

Pray, L. (2008b). Eukaryotic genome complexity. *Nature Education* 1(1):96.

Saccone, C., Gissi, C., Lanave, C., Larizza, A, Pesole G. and Reyes, A. (2000). Evolution of the mitochondrial genetic system: an overview. Gene. 261(1):153-9. doi: 10.1016/s0378-1119(00)00484-4.

Sadava, D., Hillis, D.M., Heller, H. C. and Berenbaum, M.R. (2011). "The Lac Operon" – 2008 in Life: The science of biology– Sinauer Associates Available from:
http://www.sumanasinc.com/webcontent/animations/content/lacoperon.html

Sharma, A. K. (1985). Chromosome structure. Perspective Report Ser 14, Golden Jubilee Publications. Indian National Science Academy 1-22.

Shuman, S., and Lima, C. D. (2004). The polynucleotide ligase and RNA capping enzyme superfamily of covalent nucleotidyltransferases. Curr Opin Struct Biol. 14(6):757-64. doi: 10.1016/j.sbi.2004.10.006.

Sinsheimer, R.L. (1970). Genetic Engineering: The modification of man. *Impact of Science on Society*, 20 (4): 279-291.

Speyer, J. F. (1965). Mutagenic DNA polymerase. Biochem Biophys Res Commun. 21(1):6-8. doi: 10.1016/0006-291x(65)90417-1.

Taanman, J. W. (1999). The mitochondrial genome: structure, transcription, translation and replication. Biochim Biophys Acta. 1410(2):103-23. doi: 10.1016/s0005-2728(98)00161-3.

Timmis, J. N., Ayliffe, M. A., Huang, C.Y., and Martin, W. (2004). Endosymbiotic gene transfer: organelle genomes forge eukaryotic chromosomes. Nat Rev Genet. 5(2):123-35. doi: 10.1038/nrg1271.

Wang, L., and Schultz, P. G. (2004). Expanding the genetic code. Angew Chem Int Ed Engl. 44(1):34-66. doi: 10.1002/anie.200460627.

Ward, A. J., and Cooper, T. A. (2010). The pathobiology of splicing. J Pathol. 220(2):152-63. doi: 10.1002/path.2649.

Watson, J.D. and Crick, F.H.C. (1953a). A structure for deoxyribose nucleic acid. Nature. 171:737–738.

Weimar, J. D., DiRusso, C. C., Delio, R., and Black, P. N. (2002). Functional role of fatty acyl-coenzyme A synthetase in the transmembrane movement and activation of exogenous long-chain fatty acids. Amino acid residues within the ATP/AMP signature motif of Escherichia coli FadD are required for enzyme activity and fatty acid transport. J Biol Chem. 277(33):29369-76. doi: 10.1074/jbc.M107022200.

Wells, J. (2011). Gene regulatory networks in embryos depend on pre-existing spatial coordinates. Abstract #347: Society for Developmental Biology Annual Meeting, Chicago.

Zhong, C., Andrews, J., and Zhang, S. (2014). Discovering non-coding RNA elements in *Drosophila* 3' untranslated regions. Int J Bioinform Res Appl. 10(4-5):479-97. doi: 10.1504/IJBRA.2014.062996.

Chapter 6

Non-Coding DNA: An Introduction to the Dark Matter of the Genome

By mapping out animal genomes, we now have a better idea of how the giraffe got its huge neck and why snakes are so long. Genome sequencing allows us to compare and contrast the DNA of different animals and work out how they evolved in their own unique ways. But in some cases, we're faced with a mystery. Some animal genomes seem to be missing certain genes, ones that appear in other similar species and must be present to keep the animals alive. These apparently missing genes have been dubbed Dark DNA.

---- Adam Hargreaves

Though the structural and functional perception of gene has gone through several refinement of ideas and evidence in course of advancement in molecular biology, the fundamental principle about genetic information of a species or an individual organism encrypted in its DNA sequence has persisted. If two individual species or organisms are phenotypically distinguishable, their DNA sequence should also be different. Such assumptions intrigued a group of aspiring molecular biologists to initiate the complete DNA sequencing of human genome, which, till date is the most ambitious and expensive project in the field of genomics, known as "Human Genome Project" or HGP, launched in 1990. It was a 13-year project, coordinated by the U.S. Department of Energy and National Institute of Health, with strategic partnership of Welcome Trust of U.K. along with institutional and academic help from Germany, France, Japan and China (DeLisi, 1988; International Human Genome Sequencing Consortium, 2001, 2004; Kolata, 15th April 2013). The primary goal of HGP was to identify the DNA variations at an individual level and find the effect of such variations to work on the diagnosis of it and invention of preventive measures for the next level to weed out genetic disorders affecting human life. The meticulous endeavor of the molecular geneticists, genetic engineers (biotechnologists) and other scientists in interdisciplinary fields, along with the help of advanced genetic engineering technologies and huge project funding (which was approximately 9 billion US dollars) helped its successful completion in 2003. It also helped with the emergence of a new branch in the applied field of science, known as "Bioinformatics."

The important objectives of HGP are as follows (Gonzaga-Jauregui et al, 2012; Vizzini, 19th March 2015; Jongeneel et al, 2017):

1. Sequencing around 3 billion base pairs which make up the human genome.

2. Building an advanced storage system that could store vast array of sequencing data as it is estimated that 3,300 books, each containing 1000 pages would be needed to physically record the data. Hence, a high speed computation with storage system was necessary.

3. Along with better storage systems, development of better data analysis would also be needed to analyze vast quantity of data accurately.

4. Identification and critical investigation of 20,000 – 25,000 genes of human DNA.

5. Sharing of relevant data and contemporary technological know-hows to deal with problems in health care, clinical pharmacology, bioinformatics etc.

There are a number of technological limitations in sequencing a long piece of DNA, hence, biotechnologists used random fragmentation of DNA molecule isolated from the cell, followed by cloning (subjected to amplify each DNA fragment prior cloning) it inside the hosts by using suitable vectors. Usually, "Bacterial Artificial Chromosome" or BAC and "Yeast Artificial Chromosome" or YAT are used as vectors in DNA sequencing[9] to

---

[9] DNA Sequencing: DNA sequencing is a unique process which helps to determine the nucleotide sequence in DNA. In larger aspects, DNA sequencing is used to elucidate the sequence of individual genes, or the clusters of genes (or operons) forming a stretch of genetic region (which might be a chromosome, or genomes of respective organism). It also helps to sequence RNAs and/or proteins (enzymes). As DNA is found to be the molecular foot-print of heredity DNA sequencing is also

coordinate inside the hosts of bacteria or yeast, respectively (Osoegawa et al, 2001). There are two experimental protocols used in DNA sequencing techniques which are as follows (Wolfsberg and Landsman, 1997; Parkinson and Blaxter, 2009):

    a.   Sequence Annotation: It involves sequencing of a whole set of genome, containing coding and non-coding sequence and distinguishing different regions of the sequence with specific functions recognized as "Sequence Annotation."

    b.   Expressed Sequence Tags or ESTs: It involves identification of genes that are expressed as RNA and known as "Expressed Sequence Tag" or ESTs.

Though the DNA fragments are sequenced by utilizing "Automated DNA sequencer," molecular geneticists observed some overlapping locations in them. So the mammoth task of aligning those sequences is done by a special computer-based program (Meehan, 28th August 2005). The next phase of the challenge, like preparing a genetic map of the genome, is managed with further corroboration of analytical evidence and genetic information to justify the critical observation on mode of action of "polymorphism of restriction endonuclease recognition sites" and the "Repetitive DNA sequence" (also known as microsatellites) and this unique functional entity has further been recognized as "application of polymorphism in repetitive DNA sequences" (Twyman, 2009).

However, at the end of the HGP in 2003, molecular geneticists revealed some unique outcomes of this project (Venter et al, 2001; International Human Genome Sequencing Consortium, 2004; Naidoo et al, 2011):

    a.   There are around 3164.7 million nucleotides in the human genome and .99 percent of the nucleotide bases are identical in all individuals in the human species.

    b.   There are around 30,000 genes in human genome and the number of bases vary in each gene. An average number of base in most of the genes is found to be around 3000, whereas, the largest gene (i.e. dystrophin) constitute 2.4 million bases.

---

used in the field of evolutionary biology to determine phylogenetic relationship of different organisms which helps to predict their possible mode of evolution.

c. At the chromosome level, gene diversity varies in different chromosomes contained in the human genome as the chromosome number: 1 contains highest number of genes (around 2968 genes) and Y chromosome contains (around 231 genes).

d. Molecular geneticists have traced 1.4 million locations of "single nucleotide polymorphism" or SNIPS which help to analyze chromosomal locations for genetic disorder-related sequences in humans.

e. The functions of almost half of the genes are yet to be deciphered.

f. Scientists observed that a large chunk of genome constitute repeated sequences. They don't have any coding ability but it gave an idea about chromosomal structure, function and evolution.

g. Molecular geneticists ascertained that a tiny part of the genome (2%) was found to encode amino acid sequences in protein.

So, the remaining 98% of the genome, apparently, which do not have any encoding potential of any amino acid sequences in protein synthesis, was considered to be "Junk DNA" or "ncDNA" (Ehret and DeHaller, 1963; Carey, 2015). Since 1960, it was speculated that a large part of the non-coding DNA perform minimal or no biological role which validates the usage of the term "Junk DNA" used for a long time (Pennisi, 2012). It appeared that the term "Junk DNA" was used by the famous genomic biologist, David Comings, in 1972, to define the structure and functions of non-coding DNA though the term was formally introduced later in the interdisciplinary research of genomics by Susumu Ohno (Ohno, 1972; Gregory, 2005). Nessa Carey's unique work, entitled "Junk DNA: A Journey through the Dark Matter of the Genome," firmly supported the idea that the non-coding sequences of genome is responsible for the formation of ribosomal RNAs and transfer RNAs which strongly complied with the earlier model of "Split-gene hypothesis." Though, using terminology like "Junk DNA" or "ncDNA" was done by randomly raising concerns in the world of biological science, Carey's (2015) work tried to harness it and use this term to refer specifically to the genes concerned with transcription of functional non-coding RNAs (ncRNA) and regulatory RNAs which were popularly recognized as RNA gene in the arena of genomics.

Though there is no specific quantitative data on ncRNA that was gleaned from HGP, empirical information (based on composite data source of transcriptomic and bioinformatics database) on a wide array of non-coding RNAs comprised: ribosomal RNA (rRNA), transfer RNA (tRNA), exRNA, micro RNA and piRNA, scaRNA, siRNA, snRNA, sonRNA and long ncRNA s (e.g. HOTAIR, Xist) etc. (Rinn and Chang, 2012; Bratkovic and Rogelj, 2014; Khan et al, 2016). Initially, these ncRNAs evolved as "useless junk product of transcription" as it never encoded proteins and hence, were known as "Junk RNA" or "ncRNA." Along with junk RNAs, there are some sequences like – transposons, retrotransposons, mobile genetic elements, Alu sequences, two types of retrotransposon repeated sequences (i.e. Long Interspersed Nuclear Elements or LINE and Short Interspersed Nuclear Elements or SINE), pseudogenes, telomeres etc. that are found to be the part of ncDNA (Franchini and Pollard, 2017). However, due to advancement in the field of genomics, scientists observed that these "Junk RNAs" are involved in regulation of gene expression in the post-transcriptional phase (Erdmann et al, 2001). Furthermore, scientists revealed that a number of ncRNAs (either of prokaryotic or eukaryotic origin) are engaged in "cellular nucleic acid targets through complementary base pairing" and control cell growth and differentiation (Erdmann et al, 2001). Though the ncRNAs are not directly taking part in the process of encoding protein, a number of ncRNA were found to regulate post-transcriptional gene expression through the action of proteins. Some regulatory RNAs also took part indirectly in protein translation as they are responsible for synthesizing peptide bonds in ribosomes (Erdmann et al, 2001). On a number of occasions, ncRNAs are found to be responsible for developmental and neurological disorders in cellular and molecular level which causes Alzheimer, Cancer and ageing, Duchenne muscular dystrophy, Myotonic muscular dystrophy, Retinitis pigmentosum etc. (Suzuki et al, 2012; Torres et al, 2014; Mathews, 2019). However, ncDNA has immense importance in elucidating the complex process of genetic interactions; understanding the changes of the heritable phenotypic changes that don't involve alteration in DNA sequence (also known as epigenetic actions), and exploring the uncharted, evolutionary regions of developmental biology of organisms.

Though the primary function of the non-coding DNA (ncDNA) is to transcribe ncRNA, it carries out a number of adjunct functions like such as transcriptional and translational regulation of amino acid coding sequences, scaffold attachment regions, centromere, and telomere regions. Though, earlier perception of "Junk DNA" has changed as the result of project ENCODE (Encyclopedia of DNA Elements), it ascertained that about 80% of the human genomic DNA is found to be biochemically functional "Junk DNA" (Doolittle, 2013). Though, such observation was refuted by a group of scientists, the estimated fraction of actual biological function of human genome is much lesser (between 8-15%) in the background researches of "comparative genomics."

However, the nuclear genome of any organism would either be measured in terms of total base-pairs present in the DNA or calculating the mass of DNA in the nucleus and which is expressed in pictograms (10^{-12} gm). It is usually expressed by dividing the average mass of a nucleotide by the total number of base pairs and it is standardized that 1 picogram represents 1000 million base pairs (Heslop-Harrison,2003). So, the amount of DNA in haploid nucleus (unreplicated) of gametophyte (n) is recognized as C-value or 1C value; and the DNA in unreplicated diploid nucleus (2n) or replicated haploid nucleus of sporophyte is expressed in terms of 2C value. Likewise, the replicated diploid nucleus of the genome of the concerned organism is expressed in terms of 4C value. So, as a standard practice, genome size of any organism is expressed by referring to its number of base-pairs in the haploid genome and is expressed either in kilobase or megabase (1 Mb= 1000000 base pair). Though there are exceptions in every walk of life, genome size is found to be more or less constant in a given species (DeSalle et al, 2005). It was noticed that the genome size in eukaryotes was larger than prokaryotes which is better in the course of evolutionary progress but such strategic advantage of larger genome shows more complex regulation of gene expression, controlling maintenance and inheritance of genomic material and phenotypic sustainability at a species level through the generations (Vanzan et al, 2017). In the eukaryotic organisms, the diversity of the genomes in terms of size vary from species to species. Scientists noticed that small-sized genome in some fungi are found to be around 10 Mb size, whereas, the size of some well-diversified angiosperms is found to be around 100000 Mb. The qualitative comparison of genome sizes between the eukaryotes comprising mammals and high group of angiosperms with lower group of eukaryotic organisms comprising yeasts and bacteria, revealed that complex group of eukaryotes have more DNA than simpler form of eukaryotes (Lam et al, 1997). So, it gives an impression that the greater size of genome in an organism is to be witnessed in terms of complexity of the organisms from simple to complex. It was observed by molecular geneticists that the genome size of fungal species of primitive nature yeast (*Saccharomyces cerevisiae*) was found to be 5 times bigger than the more advanced, bacterial species *Escherichia coli*, so the yeast species is supposed to be structurally and functionally advanced than *E.coli* but that is not the case (Lam et al, 1997). Scientists observed that the genomes of an angiospermic plant species like lilies and salamanders were found to be more than 10 time bigger than human species. Following the same logic these two species are supposed to be 10 time more advanced than human species but the genomic profile does not fit the assumptions (Cooper et al, 2000). In another instance, it was also found that a protozoan organism, *Polychaos dubium* (popularly known as *Amoeba*), is the possessor of the largest genome (constituted of 670000 Mb) which contains around 200 times more DNA the than human species (Gregory et al, 2000; Hodgkin, 2001). So, in terms of genome profile the structural, functional and evolutionary aspects of *Amoeba* should be more advanced and complex than human species. Hence, it was determined by the scientists that there is no co-relationship between the genomic configuration and organismal complexity in the biological world (Hodgkin, 2001). This paradox was solved by contemporary studies in the field of genomics which revealed that along with functional genes, the eukaryote genome carry non-coding sequence of DNA but don't code for proteins (Cooper et al, 2000). As the studies failed to establish that "less diversity in genes expressed in humans than in these other organisms" the amount of nuclear DNA hardly substantiated the phenotypic complexity of any organism (Lam et al, 1997). The genome size, as well as the amount of its constituent varies in different group of eukaryotic organisms and the amount of DNA in any genome is constituted of interspersed segments of exons (having encoding potential of proteins, via mRNA) and introns (having non-coding properties and DNA transcription end up with tRNA and sRNA formation). It is plausible that the ratio of coding-DNA and non-coding-DNA varied at an inter-specific level. It is mentioned earlier that completion of HGP project revealed that 98% of the human genome doesn't have encoding potential of protein translation (Smith et al, 2017). Therefore, it was noticed that in the human genome all introns, a large part of the exons and intergeneric DNA contribute to the non-coding sequence of DNA (Elgar and Vavouri, 2008). So, the presence of large quantity of non-coding DNA or "Junk DNA" or "ncDNA" in eukaryotic genomes and relative percentage of non-coding DNA out of the total amount of DNA in their respective genome seems to be the cardinal factor to substantiate phenotypic complexity of the organism at a species level and elucidates the evolutionary history of the concerned species (Lam et al, 1997). Hence, it was construed that it is not the large size of the genome of salamanders which could render as phenotypically more advanced or complex than human species, rather it is the presence of more number of genes in the genome of salamanders (Cooper et al, 2000). The idea of high extent of non-coding sequence of DNA (with respect to coding sequence of the concerned genome) in higher group of organisms in explaining the phenotypic complexity of the organisms may further be substantiated from the reference of the genomic profile of Japanese puffer fish (*Takifugu rbripes*). It's genomic size is around 12-13% of human genome, yet, even after having comparable number of genes, 85-90% of its DNA sequence was found to be non-coding kind (Elgar and Vavouri, 2008). Hence, it was unequivocally accepted that large volumes of non-coding sequence of DNA is a general feature of eukaryotic genome. Going further, scientists proved that the larger size of human genome (around 1000 time larger) with respect to *E.coli* is not due to the size of genes in the human genome, rather, the number of genes which is around 100,000—about 25 times higher than the number of genes found in *E.coli* (Cooper et al, 2000). This clearly indicates that variation of size in the genome of different organisms are not related to the number of genes, rather, it's more or less regulated by non-coding DNA or junk DNA, as the extent of it varies from species to species in higher group of eukaryotic organisms (Cooper et al, 2000).

However, there has been a paradigm shift in the comparative genomic studies due to a series of advancement in the last 50 years in terms of biotechnological protocol and perception in the field of molecular genetics. Also, genomics in terms of development of computer-aided programs cope with better quantitative and qualitative analysis of DNA material for a diverse array of organisms. The maiden approach in the study of genomics started much earlier than the major breakthrough of the molecular genetics—the discoveries of Francis and Creek's Double helix model of DNA in 1953. The earliest approach in empirical research on the estimation of the amount of DNA in a particular animal species started with analysis of their nuclei and cell suspension in 1940s (Boivin et al 1948; Vendrely and

Vendrely, 1948). One of the first efforts in the study of constancy of nuclear DNA at the species level among animal and plant organisms was started by a group of genome biologists which included Mirsky and Ris (1949); Swift (1950). During his pursuit on plant materials with *Tradescantia paludosa*, Swift (1950), invented the C-value index to define the DNA content of an unreplicated haploid nuclear genome. The earliest record of estimation regarding nuclear genome of a plant- *Lilium longifolium* was successfully done by Ogur et al, (1951). It took genome biologists around 20 years to successfully perform the DNA estimation of some plant at the interspecific level: (i) 40 fold DNA variation in the genome, observed among 22 diploid species belonged to Ranunculaceae (Rothfels et al, 1966) and interspecific level: (ii) 5 fold DNA variation in the genome of *Vicia* (Martin and Shanks, 1966). Though initial focus of genomic studies was finding the DNA constancy at species level but it gradually changed to exploration of genome size in different group of organisms which elucidated the taxonomic inter- and intra-relationship is closely related. The revelation of "C-value paradox" by Thomas (1971) elucidated that bigger size of genomes is not necessarily attributed to the phenotypic complexity of the organisms as genome contains a "fraction of DNA other than genes and their regulatory sequence." So, the emergence of "C-Value paradox" gave genomic studies a new direction where the structural and functional aspects of gene and genome as a whole is studied at a molecular level with the emerging concept of junk DNA to better define the biological evolution of. If C-value paradox is the first shift in paradigm in genomic research from "the studies of constancy of DNA" to "a fraction of DNA other than genes and their regulatory sequence" till 1980s, then the next shift in paradigm was assisted by "DNA sequencing" which was found to be the result of advancement in biotechnology in 1980s. The focus of the genomic studies further shifted from estimation of size of the genomes and its structural formation to defining it by deciphering molecular footprints of life. The contemporary advancement of genomic studies would help the next generation of genomic biologists to be introduced to 3800 animal genome—out of which 2500 are vertebrates (Gregory, 2001)—and a database on the genomes of 4000 plant taxa, maintained in Royal Botanic Garden, Kew, United Kingdom (http:// www. rbgkew. org. uk /cval /homepage.html).

However, the genomic biologist, Susumu Ohno (1972) utilized the "C-Value Paradox' to establish that even closely phylogenetically-related species might have a big difference in their respective genome size (Eddy, 2012). It was also observed by the genome biologists that the distinct genome sizes between two phylogenetically close-related plants was due to the presence of differential retrotransposon sequences (Piegu et al, 2006; Hawkins et al, 2006). In 1960s, the impact of mutation on population, particularly the human population, due to physical and biochemical mutagens was assessed under the principles of mutational load, which was independently proposed by J.B.S Haldane and H.J. Muller. They defined that "the mutational load is the decrease in fitness or viability, triggered by repeated occurrence of deleterious mutation" (Crow, 2000). Though Haldane and Muller opined that the overall effect of mutation on fitness of any organism or species has no relation to the deleterious effect of individual mutations, it was the result of the cumulative rate of mutation in each gamete X 2 (if / when mutants are dominant in nature). Such formulation indicates that effect of mutations at different locus of gene is different (Crow, 2000). Hence, it further clarified the assumption that mutation rate controlled by the mutational load out of a deletrious mutations put molecular control upon the functional loci of genes. Based on his meticulous studies, Ohno (1972), promulgated that the mammal genome should have less than 30,000 loci of genes under selection, as the mutational load might lower the fitness and extinction in the long run. The completion of HGP proved it to be a realistic judgement as the human genome constituted of around 20,000 genes (Ohno, 1972; Gregory, 2005). At the end of HGP, genome biologists noticed that around 8% of the human genome is made up of "Endogenous retrovirus sequences"[10] mostly in degenerated form. In their publication, titled "Selfish DNA: The ultimate parasite"; around 42% was found to be retrotransposons and 3% of it was recognized as left over DNA transposons (Orgel and Crick, 1980). There is no clear idea of the origin of the remaining half of the human genome at present and it is predicted to be derived from transposons. It was supposedly active around 200 million years ago but occurrence of a series of dispersive and overlapping mutations make it difficult to substantiate the presumption (Lander et al, 2001). Orgel and Crick (1980) were of the opinion that Junk DNA or ncDNA of any organism hardly attributed any "selective advantage" to the concerned organism (Wells, 2011). In the latter half of 70s, it was perceived that a large share of

---

[10] Endogenous retrovirus sequences: The reverse transcription process of the genome of retrovirus, yields endogenous retrovirus sequences.

the non-coding DNA in large-sized genomes, supposedly, originated as a result of amplification of transposable elements. As a sincere effort to review the earlier perception on origin of non-coding DNAs, Doolittle and Sapienza (1980) further clarified the evolutionary aspects of Non-coding DNA as "when a given DNA, or a class of DNAs, of unproven phenotypic function, can be shown to have evolved a strategy (as transposon) which ensures its genomic survival, then no other explanation for its existence is necessary." It was further interpolated that the existence of Junk DNA in any genome depends upon the cumulative effect of its rate of amplification in the non-coding DNA elements and the ultimate loss of it at a given time. In this regard, the endeavors of a famous scientist, molecular geneticist and plant breeding expert Barbara McClintock (1950), deserves a mention. She discovered the transposable elements of transposons in 1950. Transposons (TE) are a small fraction of genome and a DNA sequence that can move in a genome. TE—also recognized as a jumping gene—could trigger mutation or reverse it at a cellular level to alter the genetic profile of the cell and change the size of genome as well as its evolution. This was found to be present in all organisms (Bourque et al, 2018). On a number of occasions, TEs were found to be beneficial to their hosts but in most cases, they were dubbed as "Selfish DNA parasites" as they functionally resemble a virus that leads to a hypothetical co-evolutionary theory that TEs and virus are from a common mother stock (Villareal, 2005). Scientists observed that the hyperactivity in TEs could damage the structure and function of exons (coding genes). As a result, the organisms evolved the potential to counter the actions of TEs in eukaryotic organisms which is called RNA interference. However, TEs helped further speciation in large families even as evolution stalled the functions of TEs by converting them into introns (non-coding DNA sequence). Overall, scientists found that a huge presence of TEs in higher eukaryotes helped to form "interspersed repeats" in genome as these intersperse repeats stalled "gene conversion," promoting evolution of "Novel gene sequence" (Plasterk et al, 1999).

Pseudogenes, an important constituent of ncDNA was found to be the "dysfunctional relative of the genes" which lost its protein encoding potential during the course of evolution (Vanin, 1985). It is predicted to be the ultimate outcome of multiple mutations within a gene whose product (i.e. protein as well as enzyme) is not essential for survival of the concerned organism (Poliseno et al, 2010). Alternately, the emergence of pseudogenes derived from retrotransposon or as a result of duplication of functional genes in a genome became the "genomic fossil" as random mutations (like frameshift mutations, base substitutions or formation of stop codons etc.) either altered the promoter region of the respective gene or made the mRNA dysfunctional to carry out its protein translation (Zheng et al, 2007). It was also observed that translocation of functionally-active mitochondrial gene from nucleus to cytoplasm render the gene dysfunctional in the form of "Nuclear mitochondrial pseudogenes" or NUMT, which is found in a number of eukaryotic genomes (Lopez et al, 1994). However, it was suggested that pseudogenes could retain the molecular memory of protein encoding for several million years and the protein encoding potential of the dysfunctional pseudogenes might be revived so these could retrieve the transcription potential. Theoretically, the sequential reversal of the mutational sequence might help to get back the ancestral phenotype of the derived organism with altered phenotype (Petrov and Harti, 2000). Scientists noticed that telomeres, the tail-end of the chromosomal location, constituted repetitive DNA which gives protection against chromosomal decaying at the time of DNA replication. The scientists also observed that telomeres transcribed to "Telomeric repeat containing RNA" or TERRA yielded synthesizing enzyme telomerase involved in lengthening the tail-end of chromosomes (Cusanelli and Chartrand, 2014). Hence, the HGP revealed that protein coding sequence of the human genome constitutes a small fraction of DNA—about 1.5-2%—of the entire genome (Wells, 2011). So it's probable that the largest part of the genome—around 98-98.5%—is non-coding entity integrated with the formation of regulatory DNAs, ncRNA, introns, LINEs, SINEs, transposons, pseudogenes, telomeres and a number of fragmented sequences. The function of these segments have remained unknown till now (International Human Genome Sequencing Consortium, 2001; Wells, 2011). On the other hand the C-value paradox confirmed that there is no collinear relationship between the size of a genome and the structural and functional complexity of the organisms, rather, it's the content of the ncDNA or Junk DNA which is cardinal in expression of advanced and complex nature of phenotypic feature of organisms (Eddy, 2012). In pre-HGP regime, scientists considered human genomes contained around 100,000 genes but application of precise gene exploration strategies along with advanced biotechnological methods in DNA sequencing helped genome biologists to quantify the exact number of genes which was found to be between 20,000 to 25,000 (International Human Genome Sequencing Consortium, 2004; Pennisi, 2012).

The genome has gone through a series of shift in paradigm as genomic studies. 1950s started with the perception that the entire DNA constituents a genome and is responsible for biosynthesis of protein. This theory went a through monumental shift when the popularization of Junk DNA (Non-coding DNA) came up in 1960s and finally became a proven fact in 1972 with the substantial evidence produced by David Comings, Susumu Ohno (Ehret and DeHaller, 1963; Ohno, 1972; Gregory, 2005). The discovery of "C-Value paradox" by C.A. Thomas Jr. in 1971 helped to substantiate Ohno's affirmation about the structural and functional existence of "Junk DNA" (Eddy, 2012). The qualitative research on human genomics, by Dan Graur, in University of Houston, USA, indicated that around 10-15% fraction of human genome is capable of encoding functional protein and can be considered as functional DNA (Dockrill, 2017). The stereotypic perception about non-coding DNA since the 1970s through the 2000s tried to promote the idea that, structurally, "Junk DNA" has no function (as it does not directly take part in encoding protein synthesis). However, a number of genome biologists believed that Junk DNA supposedly played a cardinal role in regulating the actions of protein encoding genes, which was published in the "New Scientist" in 1972. It was a matter of concern to the genome biologists that the human genome carried up to 90% of Junk DNA which, apparently, has no functional role that could define the evolutionary supremacy of human species in the history of biological evolution of mammals (Dockrill, 2017). The successful completion of Human Genome Project or HGP, in 2003, helped biologists to define the intricate functionaries of Junk DNA at the primary level. Next in the annals of genomic studies on human genome, is "Encyclopedia of Human Genomes" or project ENCODE, which was undertaken and funded by National Human Genome Research Institute/National Institute of Health, Maryland, USA (NHGRI), in 2003 (Doolittle, 2013). According to Dan Graur of University of Houston, 400 million dollars were invested in this project, so they needed to present a unique scenario in research related to human genome. Surprisingly, the scientists of NHGRI claimed that around 80% of the human genome is functional which seems to be unbelievable to a number of genome biologists. The NHGRI scientists clarified their interpretation of functionality by stating that around 76% of the ncDNA sequences are subjected to transcription and claimed that around 50% of the genome was involved in synthesis of regulatory protein as well as acting like "transcription factors" (Pennisi, 2012). Most of the molecular biologists opined that mutations of DNA alter the stability of triplet codons and disrupt the encoding sequencing of amino acid in the protein chains; resulting in the formation of non-functional proteins, rather than formation of unique proteins expediting evolutionary progress. So, the harmful effect of mutations seemed to be proven beneficial. Even the mutational effect on DNA, previously considered harmful to the survival of the offspring, has very little cumulative deleterious effect. It would not allow them to attain adulthood unless evolution weeded out the harmful mutations so that the species could survive for a number of generations. Dan Graur promulgated that harmful effect of mutations could affect the important locations of coding DNA sequences or functional DNA sequences and accumulation of deleterious mutations in our genome would push humans to the brink of extinctions in a short span of time (Dockrill, 2017). He further contemplated that averting harmful effects of mutations is facilitated by Junk DNA as it acts as molecular safety valve to stall protein translation of the mutated genes and directly disrupts the harmful actions of mutation.

C-Value paradox established that the size of genome does not have any relationship with complexity of organisms and showed that the non-coding DNA sequences plays an important role in the evolution of eukaryotic organism. "GC-Content Enigma" is another unsolved issue in genomics where, in some advanced group of organisms, GC base-pair content is found to be disproportionately high with respect to AT content. Overall GC content varies from 20% to 70% in the genome of different organisms where genome biologists could not establish any direct relationship between higher content of GC with complexity of organisms in terms of structure and function (Agoni, 2013). "GC content enigma" was found to be the transitional link that helped to promote the most recent, unique discoveries of "Dark DNA" in 2017, emerging out of 1960s "Junk DNA." Contemporary genome biologists assumed that most of the currently practiced techniques are not competent in sequencing DNA and is therefore enriched with GC base pairs (Agoni, 2013). So, it is most likely that a number of genomes act on old DNA sequencing methods where stretches of DNA segment is rich with GC base-pairs and appears to be missing or non-existing (Moran et al, 2020). Those "difficult to sequence DNA" is dubbed as "Dark DNA" and thus the story of discovery of "Dark DNA" began (Hargreaves et al, 2017). One meticulous genome biologist from USA— Adam Hargreaves—along with his colleagues were involved in the DNA sequencing experiment of the genome of sand rat (*Psammomys obesus*). They undertook a DNA sequencing technique called "Shot gun method" to sequence the sand rat genome (Moran et al, 2020). After assembling millions of short DNA sequences to restructure a genome, the scientists noticed that a large stretch of genes were missing. Hargreaves and his associates (2017) drew a primary conclusion after comparing the genes of "conserved and present" phylogenetically-related species. If it rains, we can conclude that it came from the clouds, whether we see or not see the clouds play its role in causing precipitation on Earth's surface. Likewise, Hargraves and his colleagues (2017) inferred that whether they could see those genes in the genome of sand rat or not, the missing cluster existed somewhere in the genome as they traced the corresponding mRNAs encoded by the "invisible genes" or "missing genes." After an unsuccessful attempt with the conventional DNA sequencing technique, Hargreaves and his co-scientists (2017) used a more advanced DNA sequencing technique. He was able to trace the GC-base-pair- enriched DNA sequences, representing the "missing genes" after finding a stretch of DNA (having 88 "missing genes") with a higher percentage of GC base pair composition in comparison to a lower percentage of AT base pair composition (Moran et al, 2020). The genome biologists defined such high rate of insertion of GC- base-pair-enriched DNA repeatedly as "gene conversion" or "biased gene conversion" and this process was integrated to the process of gene recombination (Hargreaves et al, 2017).

Molecular geneticists observed that the genetic recombination in a stretch of DNA produces mismatched base pairs instead of conventional AT and GC base pairing. Along with mismatched repair of DNA, there are few more mutation-repairing correctional biases (e.g. insertions vs. deletions). Biases during recombination like mispairing after strand breakage etc. play an important role in driving "biased gene conversion" (Kostka et al, 2012; Agoni, 2013). However, the inherent genetical mechanism which always engaged in reverting the deleterious act of mutation, supposedly, repair the mismatched base-pairs by formation of AT, GC base-pairing formation. Hargreaves and his associates (2017) further proposed that this base-pair restoration process is biased as GC base-pair substitutes the AC base-pair or removes it from the genome to make a particular stretch of genome enriched with GC base-pairs and this process is recognized as "biased gene conversion" (Moran et al, 2020). This GC enriched stretch of DNA in the genome was later considered a hotspot for gene recombination and this phenomenon was also witnessed in other organisms (Moran et al, 2020). Hilderbrand et al (2010), hypothesized that the rate of GC→ AT mutation is much higher than the rate of mutation from AT→ GC. It was also found that the hydrogen bond between GC base-pair is 3, whereas, the hydrogen bond between AT base-pair is 2 which made GC more stable than AT combination thus resulting in decaying of AT base-pair location and gradually occupation by GC content. Consequently, the DNA with greater GC content is found to be more stable than the ones with low GC content. The presence of Junk DNA and the overbearing presence of GC content with respect to AT content might have an puzzling explanation. In higher groups of organisms, GC → AT mutations might be considered as a more appropriate process than single nucleotide substitutions. So it's assumed that the Junk DNA stall the mutation probability which was substantially reduced in GC poor genomes (Gregory et al, 2009). Though, such stretches of genome are either invisible or missing during exploration; it is nothing but the limitations of conventional DNA sequencing techniques which failed to trace the existence of the so-called "Dark DNA" (Biemontl and Vieiral, 2006). From an evolutionary perspective, the study on the Dark DNA of sand rat is important as the existence of GC base-pair-enriched location from the phylogenetically-close related species of sand rat, render the evolutionary biologist to understand that occurrence of "biased gene conversion" possibly existed prior to the evolution of their "last common ancestor" (Moran et al, 2020).

However, the contemporary world of biological science, particularly, the community of evolutionary biology is still skeptical about the structural and functional interpretation of GC-rich areas of genome in eukaryotes. Even its presence is logically ensured in large quantities in the genomes of some advanced group of eukaryotes. It needs to be substantiated further to determine whether the presence of such GC-rich regions would be beneficial for the survival of species and play an important role in evolution (Galtier et al, 2001). In the field of science, any small or big discoveries would be instantaneous in nature but acceptance of it by the community of science takes a long time, as it passes through the nodes of "alleged new paradigm" before accepting it or discard it. So, the statements published in 2018 by Adam Hargreaves certainly intrigues the mind (Moran et al, 2020). The scientific community are as inquisitive to know, appreciate or criticize his propositions.

As the proponent of the "Dark DNA" hypothesis, Hargreaves has created a buzz in the community of science. He admits, "The discovery of dark DNA is so recent that we are still trying to work out how widespread it is and whether it benefits those species that possess it. However, its very existence raises some fundamental questions about genetics and evolution. We may need to look again at how adaptation occurs at the molecular level. Controversially, dark DNA might even be a driving force of evolution." (Moran et al, 2020)

Reference

Agoni, V. (2013). DNA evolved to minimize frameshift mutations. *arXiv: Other Quantitative Biology.*

Biémont, C., and Vieira, C. (2006). Genetics: junk DNA as an evolutionary force. Nature. 443(7111): 521-4. doi: 10.1038/443521a.

Boivin. A. (1948). Les acides nucléiques dans la constitution cytologique et dans la vie bactérienne. C.R. Soc. Biol., Paris 142:1258.

Bourque, G., Burns, K.H., Gehring, M. Gorbunova, V., Seluanov, A., Hammell, M., Imbeault, M., Izsvák, Z., Levin, H.L., Macfarlan, T.S., Mager, D.L., and Feschotte, C. (2018). Ten things you should know about transposable elements. Genome Biol 19: 199. https://doi.org/10.1186/ s13059-018-1577-z

Bratkovic, T., and Rogelj, B. (2014). The many faces of small nucleolar RNAs. Biochim Biophys Acta 1839:438-43.

Carey, N. (2015). Junk DNA: A Journey through the Dark matter of genome. Icon Books Ltd. P.352.

Cooper, K.F., Mallory, M.J., Egeland, D.B., Jarnik, M., and Strich, R. (2000). Ama1p is a meiosis-specific regulator of the anaphase promoting complex/cyclosome in yeast. Proc Natl Acad Sci U S A. 97(26): 14548-53. doi: 10.1073/pnas.250351297.

Crow, J. F. (2000). The origins, patterns and implications of human spontaneous mutation. Nat Rev Genet. 1(1): 40-7. doi: 10.1038/35049558.

Cusanelli, E., and Chartrand, P. (2014). Telomeric noncoding RNA: telomeric repeat-containing RNA in telomere biology. Wiley Interdiscip. Rev. RNA 5, 407–419. doi: 10.1002/wrna.1220

DeLisi, C. (1988). The Human Genome Project: The ambitious proposal to map and decipher the complete sequence of human DNA. *American Scientist, 76*(5): 488-493. http://www.jstor.org/ stable/27855388

DeSalle, R., Gregory, T.R., and Johnston, J.S. (2005). Preparation of samples for comparative studies of arthropod chromosomes: visualization, *in situ* hybridization, and genome size estimation. Methods Enzymol 395: 460–488.

Dockrill, P. (19th January 2017). World-First in Ukraine as 'Three-Parent' baby born to an infertile couple. https://www.sciencealert.com/world-first-in- ukraine-as-three-parent-baby-born-to-an-infertile-couple

Doolittle, W. F. (2013). Is junk DNA bunk? A critique of ENCODE. Proc Natl Acad Sci U S A. 110(14): 5294-300. doi: 10.1073/pnas.1221376110.

Doolittle, W. F. and Sapienza, C. (1980). Selfish Genes, the Phenotype Paradigm and Genome Evolution. Nature. 284(5757): 601–603. doi:10. 1038/284601a0

Eddy, S. R. (2012). The C-value paradox, junk DNA and ENCODE. Curr Biol. 22(21): R898-9. doi: 10.1016/j.cub.2012.10.002.

Ehret, C. F., and DeHaller, G. (1963). Origin, development and maturation of organelles and organelle systems of the cell surface in *Paramecium*. Journ. Ultrastructure Research 9(1): 1-42.

Elgar, G., and Vavouri, T. (2008). Tuning in to the signals: noncoding sequence conservation in vertebrate genomes. Trends Genet. 24(7): 344-52. doi: 10.1016/j.tig.2008.04.005.

Erdmann, V.A., Barciszewska, M.Z., Szymanski, M., Hochbergm A., de Groot, N., and Barciszewski, J. (2001). The non-coding RNAs as riboregulators. Nucleic Acids Res. 29(1):189-93. doi: 10.1093/nar/29.1.189.

Franchini, L.F., and Pollard, K.S. (2017). Human evolution: the non-coding revolution. BMC Biol 15: 89. https://doi.org/10.1186/s12915-017-0428-9

Galtier, N. (2001). Maximum-likelihood phylogenetic analysis under a covarion- like model. Mol Biol Evol. 18(5): 866-73. doi: 10.1093/oxfordjournals. molbev.a003868.

Gregory, S. G., Connelly, J.J., Towers, A.J., Johnson, J., Biscocho, D., Markunas, C.A., Lintas, C., Abramson, R.K., Wright, H.H., Ellis, P., Langford, C.F., Worley, G., Delong, G.R., Murphy, S.K., Cuccaro, M.L., Persico, A., and Pericak-Vance, M.A. (2009). Genomic and epigenetic evidence for oxytocin receptor deficiency in autism. BMC Med. 7: 62. doi: 10.1186/1741-7015-7-62.

Gregory, T. R. (2001). Coincidence, Coevolution, or Causation? DNA Content, Cell Size, and the C-value Enigma. Biological Reviews 76(1): 65–101.

Gregory, T.R. (2005). Synergy between sequence and size in large-scale genomics. Nat Rev Genet 6: 699–708. doi:10.1038/nrg1674

Gregory, T. R., Herbert, P. D. N., and Kolasa, J. (2000). Evolutionary implications of the relationship between genome size and body size in flatworms and copepods. Heredity 84:201–208.

Gonzaga-Jauregui, C., Lupski, J.R., and Gibbs, R.A. (2012). Human genome sequencing in health and disease. Annu Rev Med. 63: 35-61. doi: 10.1146/annurev-med-051010-162644.

Hargreaves, A.D., Zhou, L., Christensen, J., Marlétaz, F., Liu, S., Li, F., Jansen, P.G., Spiga, E., Hansen, M.T., Pedersen, S.V.H., Biswas, S., Serikawa, K., Fox, B.A., Taylor, W.R., Mulley, J.F., Zhang, G., Heller, R.S., and Holland, P.W.H. (2017). Genome sequence of a diabetes-prone rodent reveals a mutation hotspot around the ParaHox gene cluster. PNAS U S A. 114(29):7677-7682. doi: 10.1073/pnas.1702930114.

Hawkins, J. S., Kim, H., Nason, J. D., Wing, R. A., and Wendel, J. F. (2006). Differential lineage-specific amplification of transposable elements is responsible for genome size variation in *Gossypium*. Genome Res. 16(10): 1252-61. doi: 10.1101/gr.5282906.

Heslop-Harrison, J. S. (2003). Planning for remodelling: Nuclear architecture, chromatin and chromosomes Trends in Plant Science. 8: 195-197. DOI: 10.1016/S1360-1385(03)00054-2

Hildebrand, M. S., Witmer, P. D., Xu, S., Newton, S. S., Kahrizi, K., Najmabadi, H., Valle, D., and Smith, R. J. H. (2010). miRNA mutations are not a common cause of deafness. *American Journal of Medical Genetics, Part A, 152*(3), 646-652. https://doi.org/10.1002/ajmg.a.33299

Hodgkin, J. (2001). What does a worm want with 20,000 genes? Genome Biol 2, comment2008.1.
https://doi.org/10.1186/gb-2001-2-11- comment2008

International Human Genome Sequencing Consortium. (2001). Initial sequencing and analysis of the human genome. *Nature* 409: 860–921. https://doi. org/10.1038/35057062

International Human Genome Sequencing Consortium. (2004). Finishing the euchromatic sequence of the human genome. Nature. 431(7011): 931-45. doi: 10.1038/nature03001.

Jongeneel, C.V., Achinike-Oduaran, O., Adebiyi, E., Adebiyi, M., Adeyemi, S., Akanle, B., Aron, S., Ashano, E., Bendou, H., Botha, G., Chimusa, E., Choudhury, A., Donthu, R., Drnevich, J., Falola, O., Fields, C.J., Hazelhurst, S., Hendry, L., Isewon, I., Khetani, R.S., Kumuthini, J., Kimuda, M.P., Magosi, L., Mainzer, L.S., Maslamoney, S., Mbiyavanga, M., Meintjes, A., Mugutso, D., Mpangase, P., Munthali, R., Nembaware, V., Ndhlovu, A., Odia, T., Okafor, A., Oladipo, O., Panji, S., Pillay, V., Rendon, G., Sengupta, D., and Mulder, N. (2017). Assessing computational genomics skills: Our experience in the H3ABioNet African bioinformatics network. PLoS Comput Biol. 13(6): e1005419. doi: 10. 1371/journal.pcbi.1005419.

Khan, S.Y., Hackett, S. F., and Riazuddin, S. A. (2016(Non-coding RNA profiling of the developing murine lens. Exp Eye Res. 145: 347-351. doi: 10.1016/j.exer.2016.01.010.

Kolata, G. (15th April 2013). Human Genome, Then and Now.
https://www. nytimes.com/2013/04/16/science/the-human-genome-project-then-and- now.html

Kostka, D., Hubisz, M.J., Siepel, A., and Pollard, K. S. (2012). The role of GC-biased gene conversion in shaping the fastest evolving regions of the human genome. Mol Biol Evol. 29(3): 1047-57. doi: 10.1093/molbev/ msr279.

Lam, T.L., Wong, R.S., and Wong, W.K. (1997). Enhancement of extracellular production of a Cellulomonas fimi exoglucanase in *Escherichia coli* by the reduction of promoter strength. Enzyme Microb Technol. 20(7): 482-8. doi: 10.1016/s0141-0229(96)00203-7.

Lander, E.S., Linton, L.M., Birren, B., Nusbaum, C., Zody, M.C., Baldwin, J., Devon, K., Dewar, K., Doyle, M., FitzHugh, W., Funke, R., Gage, D., Harris, K., Heaford, A., Howland, J., Kann, L., Lehoczky, J., LeVine, R., McEwan, P., McKernan, K., Meldrim, J., Mesirov, J.P., Miranda, C., Morris, W., Naylor, J., Raymond, C., Rosetti, M., Santos, R., Sheridan, A., Sougnez, C., Stange-Thomann, Y., Stojanovic, N., Subramanian, A., Wyman, D., Rogers, J., Sulston, J., Ainscough, R., Beck, S., Bentley, D.,

Burton, J, C.lee, C., Carter, N., Coulson, A., Deadman, R., Deloukas, P., Dunham, A., Dunham, I., Durbin, R., French, L., Grafham, D., Gregory, S., Hubbard, T., Humphray, S., Hunt, A., Jones, M., Lloyd, C., McMurray, A., Matthews, L., Mercer, S., Milne, S., Mullikin, J.C., Mungall, A., Plumb, R., Ross, M., Shownkeen, R., Sims, S., Waterston, R.H., Wilson, R.K., Hillier, L.W., McPherson, J.D., Marra, M.A., Mardis, E.R., Fulton, L.A., Chinwalla, A.T., Pepin, K.H., Gish, W.R,. Chissoe, S.L., Wendl, M.C., Delehaunty, K.D., Miner, T.L., Delehaunty, A., Kramer, J.B., Cook, L.L., Fulton, R.S., Johnson, D.L., Minx, P.J., Clifton, S.W., Hawkins, T., Branscomb, E., Predki, P., Richardson, P., Wenning, S., Slezak, T., Doggett, N., Cheng, J.F., Olsen, A., Lucas, S., Elkin, C., Uberbacher, E., Frazier, M., Gibbs, R.A., Muzny, D.M., Scherer, S.E., Bouck, J.B., Sodergren, E.J., Worley, K.C., Rives, C.M., Gorrell, J.H., Metzker, M.L., Naylor, S.L., Kucherlapati, R.S., Nelson, D.L., Weinstock, G.M., Sakaki, Y., Fujiyama, A., Hattori, M., Yada, T., Toyoda, A., Itoh, T., Kawagoe, C., Watanabe, H., Totoki, Y., Taylor, T., Weissenbach, J., Heilig, R., Saurin, W., Artiguenave, F., Brottier, P., Bruls, T., Pelletier, E., Robert, C., Wincker, P., Smith, D.R., Doucette-Stamm, L., Rubenfield, M., Weinstock, K., Lee, H.M., Dubois, J., Rosenthal, A., Platzer, M., Nyakatura, G., Taudien, S., Rump, A., Yang, H., Yu, J., Wang, J., Huang, G., Gu, J., Hood, L., Rowen, L., Madan, A., Qin, S., Davis, R.W., Federspiel, N.A., Abola, A.P., Proctor, M.J., Myers, R.M., Schmutz, J., Dickson, M., Grimwood, J., Cox, D.R., Olson, M.V., Kaul, R., Raymond, C., Shimizu, N., Kawasaki, K., Minoshima, S., Evans, G.A., Athanasiou, M., Schultz, R., Roe, B.A., Chen, F., Pan, H., Ramser, J., Lehrach, H., Reinhardt, R., McCombie, W.R., de la Bastide, M., Dedhia, N., Blöcker, H., Hornischer, K., Nordsiek, G., Agarwala, R., Aravind, L., Bailey, J.A., Bateman, A., Batzoglou, S., Birney, E., Bork, P., Brown, D.G., Burge, C.B., Cerutti, L., Chen, H.C., Church, D., Clamp, M., Copley, R.R., Doerks, T., Eddy, S.R., Eichler, E.E., Furey, T.S., Galagan, J., Gilbert, J.G., Harmon, C., Hayashizaki, Y., Haussler, D., Hermjakob, H., Hokamp, K., Jang, W., Johnson, L.S., Jones, T.A., Kasif, S., Kaspryzk, A., Kennedy, S., Kent, W.J., Kitts, P., Koonin, E.V., Korf, I., Kulp, D., Lancet, D., Lowe, T.M., McLysaght, A., Mikkelsen, T., Moran, J.V., Mulder, N., Pollara, V.J., Ponting, C.P., Schuler, G., Schultz, J., Slater, G., Smit, A.F., Stupka, E., Szustakowski, J., Thierry-Mieg, D., Thierry-Mieg, J., Wagner, L., Wallis, J., Wheeler, R., Williams, A., Wolf, Y.I., Wolfe, K.H., Yang, S.P., Yeh, R.F., Collins, F., Guyer, M.S., Peterson, J., Felsenfeld, A., Wetterstrand, K.A., Patrinos, A., Morgan, M.J., de Jong, P., Catanese, J.J., Osoegawa, K., Shizuya, H., Choi, S., Chen, Y.J., and Szustakowki, J. (2001). International Human Genome Sequencing Consortium. Initial sequencing and analysis of the human genome. Nature. 409(6822): 860-921. doi: 10.1038/35057062. Erratum in: Nature 2001 Aug 2;412(6846):565. Erratum in: Nature 2001 Jun 7;411(6838):720. Szustakowki, J [corrected to Szustakowski, J].

Martin, P., and Shanks, R. (1966). Does *Vicia faba* have Multi-stranded Chromosomes? Nature 211: 650–651 (1966). https://doi.org/10. 1038/211650a0

Mathews, D. H. (2019). How to Benchmark RNA Secondary Structure Prediction Accuracy. Methods. 162-163: 60-67.

McClintock, B. (1950) The origin and behavior of mutable loci in maize. PNAS USA 36(6):344–355.

Meehan, A. (28th August 2005). How it works? Automated DNA sequencer. https://www.the-scientist.com/how-it-works/how-it-works-automated-dna-sequencer-48443

Mirsky, A. E., and Ris, H. (1949). Variable and constant components of chromosomes. Nature. 163(4148): 666. doi: 10.1038/163666a0.

Moran, R.L., Catchen, J.M., and Fuller, R.C. (2020). Genomic Resources for Darters (Percidae: Etheostominae) Provide Insight into Postzygotic Barriers Implicated in Speciation, *Molecular Biology and Evolution,* 37(3):711–729, https://doi.org/10.1093/molbev/msz260

Naidoo, N., Pawitan, Y., Soong, R., Cooper, D.N., and Ku, C. S. (2011). Human genetics and genomics a decade after the release of the draft sequence of the human genome. Hum Genomics. 5(6): 577-622. doi: 10.1186/1479- 7364-5-6-577.

Ogur, M., Erickson, R.O., Rosen, G.U., Sax, K.B., and Holden, C. (1951). Nucleic acids in relation to cell division in *Lilium longiflorum.* Experimental Cell Research. 2:73–89.

Ohno, S. (1972). So much "junk" DNA in our genome. Brookhaven Symp. Biol. 23: 366-70.

Orgel, L.E. and Crick, F.H. (1980). Selfish DNA: the Ultimate Parasite, Nature. 284(5757): 604–607. doi:10.1038/284604a0

Osoegawa, K., Mammoser, A.G., Wu, C., Frengen, E., Zeng, C., Catanese, J.J., and de Jong, P.J. (2001). A bacterial artificial chromosome library for sequencing the complete human genome. Genome Res. 11(3):483-96. doi: 10.1101/gr.169601.

Parkinson, J., and Blaxter, M. (2009). Expressed sequence tags: an overview. Methods Mol Biol. 533: 1-12. doi: 10.1007/978-1-60327-136-3_1.

Pennisi, E. (2012). Genomics. ENCODE project writes eulogy for junk DNA. Science. 337(6099):1159-1161. doi: 10.1126/science.337.6099.1159.

Petrov, D.A., and Hartl, D. L. (2000). Pseudogene evolution and natural selection for a compact genome. J Hered. 91(3): 221-7. doi: 10. 1093/jhered/91.3.221.

Piegu, B., Guyot, R., Picault, N., Roulin, A., Sanyal, A., Kim, H., Collura, K., Brar, D.S., Jackson, S., Wing, R.A., and Panaud, O. (2006). Doubling genome size without polyploidization: dynamics of retrotransposition-driven genomic expansions in Oryza australiensis, a wild relative of rice. Genome Res. 16(10): 1262-9. doi: 10.1101/gr.5290206. Epub 2006 Sep 8.

Erratum in: Genome Res. 2011 Jul; 21(7):1201. Saniyal, Abhijit [corrected to Sanyal, Abhijit].

Plasterk, R.H., Izsvák, Z., and Ivics, Z. (1999). Resident aliens: the Tc1/mariner superfamily of transposable elements. Trends Genet. 15(8): 326-32. doi: 10.1016/s0168-9525(99)01777-1.

Poliseno, L., Salmena, L., Zhang, J., Carver, B., Haveman, W.J., and Pandolfi, P. P. (2010). A coding-independent function of gene and pseudogene mRNAs regulates tumour biology. Nature. 465(7301): 1033-8. doi: 10.1038/nature09144.

Rinn, J.L., and Chang, H.Y. (2012). Genome regulation by long noncoding RNAs. Annu Rev Biochem. 81:145-66. doi: 10.1146/annurev-biochem-051410-092902.

Rothfels, K., Sexsmith, E., Heimburger, M., and Krause, M.O. (1966). Chromosome size and DNA content of species of anemone L. and related genera (Ranunculaceae). Chromosoma 20: 54–74 (1966). https://doi.org /10.1007/BF00331898

Smith, T., Heger, A., and Sudbery, I. (2017). UMI-tools: modeling sequencing errors in Unique Molecular Identifiers to improve quantification accuracy. Genome Res. 27(3): 491-499. doi: 10.1101/gr.209601.116.

Suzuki, T., Sugiyama, M., Wakazono, K., Kaneko, Y., and Harashima, S. (2012). Lactic-acid stress causes vacuolar fragmentation and impairs intracellular amino-acid homeostasis in Saccharomyces cerevisiae. J Biosci Bioeng. 113(4): 421-30. doi: 10.1016/j.jbiosc.2011.11.010.

Swift, H. (1950). The constancy of desoxyribose nucleic acid in plant nuclei. PNAS USA 36: 643.

Thomas, C.A. Jr. (1971). The genetic organization of chromosomes. Annu. Rev. Genet. 5: 237–256.

Torres, R., Martin, M.C., Garcia, A., Cigudosa, J.C., Ramirez, J.C., and Rodriguez-Perales, S. (2014). Engineering human tumour-associated chromosomal translocations with the RNA-guided CRISPR-Cas9 system. Nat Commun. 5: 3964. doi: 10.1038/ncomms4964.

Twyman, R. M. (2009). Transcriptional Silencing. In: Larry R. Squire (Ed.) Encyclopedia of Neuroscience. Academic Press. Pp. 1099-1104.

Vanin, E.F. (1985). Processed pseudogenes: characteristics and evolution. Annu Rev Genet. 19:253-72. doi: 10.1146/annurev.ge.19.120185.001345.

Vanzan, L., Sklias, A., Herceg, Z., and Murr, R. (2017). Mechanisms of Histone modifications. (Ed. Trygve O. Tollefsbol). In. Handbook of Epigenetics. Academic Press. Pp 25-46. https://doi.org/10.1016/B978-0-12-805388- 1.00003-1.

Vendrely, R., and Vendrely, C. (1948). La teneur du noyau cellulaire en acide désoxyribonucléique à travers les organes, les individus et les espèces animales [The content of dexyxyconucleic acid in the cell nucleus through organs, individuals and animal species]. Experientia. 4(11): 434-6. French. doi: 10.1007/BF02144998.

Venter, J.C., Adams, M.D., Myers, E.W., Li, P.W., Mural, R.J., Sutton, G.G., Smith, H.O., Yandell, M., Evans, C.A., Holt, R.A., Gocayne, J.D., Amanatides, P., Ballew, R.M., Huson, D.H., Wortman, J.R., Zhang, Q., Kodira, C.D., Zheng, X.H., Chen, L., Skupski, M., Subramanian, G., Thomas, P.D., Zhang, J., Gabor Miklos, G.L., Nelson, C., Broder, S., Clark, A.G., Nadeau, J., McKusick, V.A., Zinder, N., Levine, A.J., Roberts, R.J., Simon, M., Slayman, C., Hunkapiller, M., Bolanos, R., Delcher, A., Dew, I., Fasulo, D., Flanigan, M., Florea, L., Halpern, A., Hannenhalli, S., Kravitz, S., Levy, S., Mobarry, C., Reinert, K., Remington, K., Abu-Threideh, J., Beasley, E., Biddick, K., Bonazzi, V., Brandon, R., Cargill, M., Chandramouliswaran, I., Charlab, R., Chaturvedi, K., Deng, Z., Di Francesco, V., Dunn, P., Eilbeck, K., Evangelista, C., Gabrielian, A.E., Gan, W., Ge, W., Gong, F., Gu ,Z., Guan, P., Heiman, T.J., Higgins, M.E., Ji, R.R., Ke, Z., Ketchum, K.A., Lai, Z., Lei, Y., Li, Z., Li, J., Liang, Y., Lin, X., Lu, F., Merkulov, G.V., Milshina, N., Moore, H.M., Naik, A.K., Narayan, V.A., Neelam, B., Nusskern, D., Rusch, D.B., Salzberg, S., Shao, W., Shue, B., Sun, J., Wang, Z., Wang, A., Wang, X., Wang, J., Wei, M., Wides, R., Xiao, C., Yan, C., Yao, A., Ye, J., Zhan, M., Zhang, W., Zhang, H., Zhao, Q., Zheng, L., Zhong, F., Zhong, W., Zhu, S., Zhao, S., Gilbert, D., Baumhueter, S., Spier, G., Carter, C., Cravchik, A., Woodage, T., Ali, F., An, H., Awe, A., Baldwin, D., Baden, H., Barnstead, M., Barrow, I., Beeson, K., Busam, D., Carver, A., Center, A., Cheng, M.L., Curry, L., Danaher, S., Davenport, L., Desilets, R., Dietz, S., Dodson, K., Doup, L., Ferriera, S., Garg, N., Gluecksmann, A., Hart, B., Haynes, J., Haynes, C., Heiner, C., Hladun, S., Hostin, D., Houck, J., Howland, T., Ibegwam, C., Johnson, J., Kalush, F., Kline, L., Koduru, S., Love, A., Mann, F., May, D., McCawley, S., McIntosh, T., McMullen, I., Moy, M., Moy, L., Murphy, B., Nelson, K., Pfannkoch, C., Pratts, E., Puri, V., Qureshi, H., Reardon, M., Rodriguez, R., Rogers, Y.H., Romblad, D., Ruhfel, B., Scott, R., Sitter, C., Smallwood, M., Stewart, E., Strong, R., Suh, E., Thomas, R., Tint, N.N., Tse, S., Vech, C., Wang, G., Wetter, J., Williams, S., Williams, M., Windsor, S., Winn-Deen, E., Wolfe, K., Zaveri, J., Zaveri, K., Abril, J.F., Guigó, R., Campbell, M.J., Sjolander, K.V., Karlak, B., Kejariwal, A., Mi, H., Lazareva, B., Hatton, T., Narechania, A., Diemer, K., Muruganujan, A., Guo, N., Sato, S., Bafna, V., Istrail, S., Lippert, R., Schwartz, R., Walenz, B., Yooseph, S., Allen, D., Basu, A., Baxendale, J., Blick, L., Caminha, M., Carnes-Stine, J., Caulk, P., Chiang, Y.H., Coyne, M., Dahlke, C., Mays, A., Dombroski, M., Donnelly, M., Ely, D., Esparham, S., Fosler, C., Gire, H., Glanowski, S., Glasser, K., Glodek, A., Gorokhov, M., Graham, K., Gropman, B., Harris, M., Heil, J., Henderson, S., Hoover, J., Jennings, D., Jordan, C., Jordan, J., Kasha, J., Kagan, L., Kraft, C., Levitsky, A., Lewis, M., Liu, X., Lopez, J., Ma, D., Majoros, W., McDaniel, J., Murphy, S., Newman, M., Nguyen, T., Nguyen, N., Nodell, M., Pan, S., Peck, J., Peterson. M., Rowe, W., Sanders, R., Scott, J., Simpson, M., Smith, T., Sprague, A., Stockwell, T., Turner, R., Venter, E., Wang, M., Wen, M., Wu,

D., Wu, M., Xia, A., Zandieh, A., and Zhu, X. (2001). The sequence of the human genome. Science. 291(5507): 1304-51. doi: 10.1126/science.1058040. Erratum in: Science 2001 Jun 5;292(5523):1838.

Villarreal, L.P. (2005).Viruses and The Evolution of Life Washington: ASM Press. doi: 10.1128/9781555817626

Vizzini, C. (19th March 2015). The human variome project. https://www.sciencediplomacy.org/article/2015/human-variome-project

Wells, J. (2011). The Myth of Junk DNA. Discovery Institute Press, Seattle.

Wolfsberg, T.G., and Landsman, D. (1997). A comparison of expressed sequence tags (ESTs) to human genomic sequences. Nucleic Acids Res. 25(8): 1626-32. doi: 10.1093/nar/25.8.1626.

Zheng, D., Frankish, A., Baertsch, R., Kapranov, P., Reymond, A., Choo, S.W., Lu, Y., Denoeud, F., Antonarakis, S.E., Snyder, M., Ruan, Y., Wei, C.L., Gingeras, T.R., Guigó, R., Harrow, J., and Gerstein, M.B. (2007). Pseudogenes in the ENCODE regions: consensus annotation, analysis of transcription, and evolution. Genome Res. 17(6): 839-51. doi: 10.1101/gr.5586307.

Chapter 7

In Quest of Hidden Genes:
Exploring the Junkyard of Genome

The behavior of some visible objects such as galaxies is affected more abundant, invisible "dark matter" and "dark energy." The analogy with genetics is that for decades, because of the simplicity of the genetic code, we biologists have been able to see the "stars" of the genome, to see exactly where genes are encoded in DNA...the genes that we see occupy just a small fraction of DNA. A much larger part of our DNA consists of sequences that are not part of the simple code for any gene and whose function cannot be deciphered simply by reading the genome. This is the "dark matter" of the genome. Just as dark matter in the universe governs the behavior of visible bodies, the dark matter of our DNA controls where and when genes are used for development.

---- Sean Carroll

Is there a simple understanding of existence of any entity continuously? Perhaps not. It is not necessary that whatever we see with our eyes exists, like the blue sky, which is nothing but space of our galactic system. Since our childhood, we have seen the twinkling stars which have left pleasant memories in our minds but logically, whatever we witnessed was nothing but the light from those starlets which traversed millions of light years in the form of high velocity particle travelling at the rate of 1,86,000 miles/sec to reach us. On the other hand, it is not also necessary that we see everything that exists around us, like dark matter in the universe. Findings by astrophysicists, astronomers, and cosmologists have revealed that the universe constitutes 80% dark matter that we don't see but its gravitational effect along with various astrophysical experiments and evaluations prove its existence. Sean Carroll's astrophysical analogy of "dark matter" of the universe to the "dark matter" of genome seems to be unique but it's not seemingly well-defined. When astronomers and cosmologists tried to decipher the "dark matter" in the universe with a series of theories and hypothesis like the principle of Quantum Physics they were able to define the identity of the "astrophysical identity of the dark matter." Similarly, contemporary advancement of molecular biology and genomics helped scientists to reach the root of "Dark matter" of genome as they could easily explore its base constituents—DNA. Once aware of the ultimate destination, it's a process of trial and error of judging, defining and redefining an object or principle from direct perspectives which would reveal the ultimate resolution. In this situation, the resolution of the identity of the "dark matter of genome" has more strategic advantage than its astronomical counterpart (Mohlenhoff et al, 2005).

The term "Dark DNA" has gone through a series of diverse names, such as "Selfish DNA," "non-coding DNA," "non-functional DNA," "junk DNA" etc. since the 1960s. In the era of cytological study of genes, whenever there was development in the advancement of histochemical researches along with the advancement in microscopic techniques, it helped the cyto-geneticists to study the different locations of chromosomal segment—through color staining—and understand the structural and functional identity of gene. The distinct part of the chromosome with the darker stain (recognized as heterochromatin) was found to be functionally less active in comparison to the lighter-stained euchromatin segment. The dark stained segments or bands of chromosome were initially recognized as dark regions of genome (Swift et al, 1964). However, the term "dark DNA" was initiated by Yamashita and Kamen, (1968) when they used this terminology instead of DNA repairing against mutation-induced damage. It was observed that the most of the organisms evolved with inherent, well-devised physical, chemical and behavioral safe-guards that protect DNA from deleterious effect of mutation and repaired it from mutation. There are two basic and distinct repair methods:

a. Photo-enzymatic repair: involved in repairing light DNA.

b. Dark DNA repair: responsible for nucleotide excision repair.

Hence, Yamishita and Kamen (1968) used the term "Dark DNA" to mean the "nucleotide excision repair" of DNA. Intrigued by the preliminary findings of the cytogeneticists in sixties (Swift et al, 1964) researchers of the late seventies and early eighties further identified the darker-stained heterochromatin and the lighter-stained euchromatin, observed in the polytene chromosomes of the salivary glands in *Drosophila melanogaster* (fruit fly), as dark genetic material (Roberts, 1975). The term "heterochromatin" was coined by Heitz in 1928 and it has been used to refer to the morphologically condensed state of chromatin (Ris and Kubai, 1970; Sharma et al, 1986). After the discovery of euchromatin and heterochromatin, cyto-geneticists observed that neither duplication, deletions nor spontaneous mutation-induced alteration in heterochromatin or heterochromatic segment of the chromosomes, made any phenotypic changes to the concerned organism. The heterochromatin never took part in the crossing-over (chiasma formation) process so its involvement in genetic recombination was less likely, which led the scientists to consider the dark-stained part of the chromatin as a genetically inert region of genome (Sharma et al, 1986). Based on the distinct cytogenetical staining difference at the preliminary level of genetical functions, it was unethical to conclude with certainty that the dispersed chromatin or euchromatin is a genetically active region, whereas, the heterochromatin or the condensed chromatin region is genetically inert. However, the discovery paved the way for Dark DNA studies during the seventies (Muller, 1928; Evans et al, 1973; Comings and Okada, 1974).

In 2003, there was a big twist in the field of explorations of Dark DNA. The scientists of ENCODE (Encyclopedia of DNA Elements) consortium, which is the follow-up project of HGP, funded by National Human Genome Research Institute in United States, tried to redefine the functional mode of DNA, which in turn, led to redefining "Dark DNA" as "functional DNA." According to the interpretation of scientists in the ENCODE consortium, a transcribed DNA sequence would be considered "functional" and such liberal perception of function such as encoding protein or protein translation was not prioritized in their consideration of "functional behavior" (White et al, 2013) of DNA. This would be equivalent to a human individual who would be considered functional if they eat, move, breathe, grow and reproduce, without considering their behavioral traits such as communicating expression or even whether they walk on two legs or use two hands. As a result of this shift in perception, ENCODE claimed that 80% of human genome are "functional" (The ENCODE Project Consortium, 2012). Consequently, such unique interpretation of "functional behavior" of the lion's share of human genome was instantaneously picked up and publicized by popular science forums, digital media and the press to uphold the enigmatic characteristics of "functional DNA." It further led to an outcry among the denialists of Dark DNA. Contemporary scientists, who maintain a strategic liaison between genome biology and evolutionary biology, refuted the phenomenon of "death of Dark DNA" as they contended that the transcription and splicing witnessed during biochemical testing in the human genome should be treated as an indicator of genetic function rather than genomic functions. This is because genomic functions are integrated to genomic conservation (a part of comparative genomics) and appraisal of genomic conservation is a tedious process to deal with even among closely-related species due to diversity of genome sizes (Kellis et al, 2014). They further affirmed that ENCODE could successfully streamline the concept of "bio-medically functional elements of genome" rather than elucidating the "evolutionary functional elements" as an evolutionary selection has no potential to define any function as part of a genome at a molecular level (Germain et al, 2014).

After decades of debate, the concept of Dark DNA materialized with certainty in the 1990s when an earlier group of molecular geneticists, discretely referred to the DNA molecules witnessed as the dark chromosomal bands (observed in the interphase state of chromosome, stained with Giesma stains) (Manuelidis, 1990). The 13 years of Human Genome Project—from 1990 to 2003—narrowed down the perception of non-coding sequences of DNA or Junk DNA, in terms of structure and function, of the 98% of the DNA sequence of the human genome that does not encode the protein translation. Basically, the successful completion of HGP in 2003, paved the way for future genomicists and molecular geneticists to investigate the identity and function of Junk DNA as well as exploring the molecular footprints of Dark DNA. It is also observed that scientific literature (in form of books and digital media) pertaining to the interdisciplinary field of molecular genetics, evolutionary biology, genomics, biotechnology, bioinformatics etc., has referred to "Dark DNA" quite frequently, to support a group of readers of academics, non-academics, researchers, students and enthusiastic readers of molecular evolution and biological discoveries, as a way to share ideas and theories of recent advancement related to the role of "Dark DNA" in relation to biological evolution of life. Though, patronization of any scientific principle or discovery or theory is not a fitting attitude to understand the scientific interpretation of cause and effect of any entity and its functions, yet, the community of science believes in evolutionary doctrine which has showed the way in the quest of molecular footprints of life. They have started leaning towards the "principle of Dark DNA" and familiarize themselves with new perspectives of biological evolution in the 21st century. Sean Carroll's eloquent interpretation of structure and functions of "non-coding Dark DNA" clearly elucidates the existence of non-coding sequence of DNA in the genome. In Carroll's exemplary work "Endless Forms Most Beautiful: The New Science of Evo Devo and the making of the Animal Kingdom," he used the analogy of "Dark Matter" in the universe to explain the "Dark DNA" of genetics. Dark Matter is thought to consist of 80% of the Universe, yet it is invisible. Only its gravitational properties prove its existence. Similarly, defending the structural interpretation of the Dark DNA, Carrol (2005), Mohlenhoff et al (2005) stated that the existence of some DNA in the nuclei rendered the chromatin visibly opaque. This further caused absence of "DNA signals" in the backdrop of infrared light. Such apparent non-existence of DNA could not have been ascertained if the concerned nuclei was devoid of DNA molecules.

Since 1977, the simple discovery of DNA sequencing (Maxam and Gilbert, 1977; Sanger et al, 1977) was an enormous leap in terms of development of instruments, advancement of techniques and methodologies like Polymerase Chain reactions or PCR technologies. The advancement and investigation in all the fields of molecular genetics and genomics in the last three decades made DNA sequencing techniques like PGM technique, AmpliSeq Technology etc. competent enough to sequence very small DNA molecules (Kieleczawa, 2006). Yet, sequencing of certain parts was found to be difficult for molecular biologists, either due to complexity in structure or if the DNA template is found to be of poor quality available in small quantity in parts of larger complex nature of genome, or contain locations difficult for DNA sequencing (Adams et al, 2000). Though there are a number of factors to be considered, the GC-enriched region of the genome is found to be the most difficult region for DNA sequencing (Kieleczawa, 2006). As it's difficult to determine the upper and lower content of GC, we could consider a DNA template that is difficult for DNA sequencing. Scientists have found that a DNA template of 100-150 base-long-part of the DNA, having 60-65% GC-enriched base sequence, is very difficult to sequence (Kieleczawa, 2005, 2006).

Finally, when Adam Hargreaves and his associates, in 2017, engaged in sequencing the diabetes-prone sand rat (*Psammomys obeus*), they noticed that a number of DNA stretches in their final form were visually missing.

The identity of those missing stretches of DNA sequences was recognized as "Dark DNA." Hargreaves (25th August 2017), further ascertained that sequencing of the DNA templates of the genome extracted from the sand rat was difficult. After biochemical examination, scientists confirmed that the stretches of Dark DNA constitute of GC-enriched base sequence. In their publication, titled "Genome sequence of a diabetes-prone rodent reveals a mutation hotspot around the ParaHox gene cluster," which was published in the 114th volume of *PNAS*, Hargreaves and his associates (2017) clearly stated:

> *These transcripts show unusually high GC content in most cases, indicating that a large contiguous stretch of elevated GC had either been underrepresented in initial sequencing data or had failed to assemble correctly, most likely due to nucleotide compositional bias. We term such cryptic or hidden sequence 'dark DNA.'*

The structural identity and existence of "non-coding DNA sequences" or Junk DNA in the genome of eukaryotes was unable to encode protein. Apart from these basic ideas, Hargreaves and his colleagues shared their observations on the evolutionary functions of the genome. He stated that the genome was regulated by its constituent Dark DNA, enriched with GC base pairs in relation to its structural evolution and the potential role in biological evolution of the organisms at the species level or higher. Scientists observed that not only mammals such as sand rats or human species exhibited this phenomenon but this was also evident in avian species, where 15% of genes are considered to be "missing" due to their integration in GC-enriched location within the stretches of avian genome (Hron et al, 2015). Going back to square one to figure out why a part of genomic sequences of sand rat were difficult for DNA sequencing, the scientists observed that these apparently difficult "DNA sequence to be part of genome," are biochemically GC-rich locations (Hargreaves et al, 2017). So, the next concern was what makes a stretch of DNA rich with GC base pairs. Genome biologists generally observed that faster rate of evolution of a genomic region of a species, with respect to its homologs of genomes of its related species appeared as GC-rich region may be triggered by a number of factors:

a. On the basis of the mutational impact on the stretches of DNA in certain locations of genome, scientists traced the regions of genome with high rate of mutation as "hotspots" and the region identified with low mutation rates were recognized as "coldspots." A number of scientists hypothesized that high rate of base-pair substitutions (GC in place of AT) in the Dark DNA supposedly expedited due to hyperactive mode of "hotspots."

b. The other group of scientists hypothesized that high rate of nucleotide substitution was supported by positive or Darwinian selection or by relaxation of selection. As Hargreaves et al (2017) precluded the possibility of "positive selection" in faster evolution of GC-rich segments of sand rat genome which left the possibility like antagonistic interaction of high rate of mutations and relaxation of selection which was supposed to render the increased rate of mutation (changing the base pairs) in a sustainable range.

c. A small group of molecular geneticists came up with their hypothesis of "biased mutation pressure" which defined that occurrence of some mutations at the particular locations (DNA stretches with desired base-pair rich locations) of genome happens at a higher extent than other mutations. The hypothesis of biased mutations could explain the GC-rich region of the DNA stretch of bacterial genome but it does not explain the occurrence of GC-rich region of Dark DNA in eukaryotes. Though, this hypothesis was further supported with little modifications, a number of genomicists opined that preponderance of GC-rich region of Dark DNA seems to be the result of GC-biased gene conversion (Escobar et al, 2011).

However, as far as the well-researched facts and findings of the genome biologists on the "non-coding sequences of DNA" as well as non-functional region of genome is concerned, the interpretation of the structural entities of "Dark DNA" or "Selfish DNA" was an elucidated narrative of molecular foot-print of life, while its contemporary functional interpretation which helped to understand the functional role of Dark DNA in the molecular evolution of eukaryotes remained poorly defined. Hargreaves (2018) have taken a bold step to understand the evolutionary functions of Dark DNA in the backdrop of the classical Darwinian principle of natural selection and the contemporary interpretations of molecular evolution theorised by Nei Masatoshi (2013) in "Mutation driven evolution." More recently, Hargreaves (2018) presented an eloquent narration on the evolutionary function of Dark DNA where he stated, *"Most textbooks describe evolution as a two-step process. First, a steady trickle of random genetic mutation creates variation in an organism's DNA. Then, natural selection acts like a filter, deciding which mutations are passed on. This usually depends on whether they confer some sort of advantage, although not everything produced over the course of evolution is an adaptation. So, natural selection is the sole driving force pushing the direction in which organisms evolve. But add dark DNA to the picture, and that's not necessarily the case. If genes contained within these mutation hotspots have a greater chance of mutating than those elsewhere, they will display more variation on which natural selection can act, so the traits they confer will evolve faster. In other words, dark DNA could influence the direction of evolution, giving a driving role of mutation. Indeed, my colleagues and I have suggested that mutation rates in dark DNA may be so rapid that natural selection cannot act fast enough to remove deleterious variants in the usual way. Such genes might even become adaptive later on if a species faces a new environmental challenge.*" Hargreaves et al (2017) have hypothetically defined that rate of mutation in the Dark DNA seems to be very high and that natural selection would not weed out the "deleterious variants" efficiently. If we go through the principles of "Mutation driven evolution," it clearly states that negative impact of mutation is witnessed among the smaller size of population (less number of mutated individuals, in respect to non-mutated individuals) and of low fitness. So, evolutionary biologists did not accept the explanation of Hargreaves et (2017) as logical as natural selection could bring down the fitness but it does not have any role in weeding-out the deleterious variants evolving as a result of recurrent or spontaneous mutation. Hence, the functional role of Dark DNA in regulating evolution of eukaryotic organisms was not satisfactorily defined.

The existence of a large part of identical sequence of "Dark DNA" in the genomes of different eukaryotic including human species, with no substantial functions, promoted the scientists to solve functional riddle of DNA sequences which has no potential for encoding protein. In 2004, Gill Bejerano, the famous genomicist at Stanford University, USA, and his colleagues discovered 481 segments of DNA (around 200 base-pairs long) that looked identical and were found in the genomes of different vertebrates like mice, rats, chickens, dogs and humans. These were recognized as "Ultraconserved Elements," though they did not have any protein encoding potential. It showed little changes in the next generation, which Bejerano and his colleagues (2004) intrigued as they tried to find two main enigmatic issues to clarify:

a. All those vertebrate species (like chickens, rats humans etc.) evolved through independent lineages of evolution in the last 200 million years but what made these diverse array of species possess such "ultraconserved elements" of DNA in their genomes?

b. While DNA mutation is a normal phenomenon, the mutation of the DNA configuration/profile of any organism from one generation to the next appeared with little changes. Then what made those "ultraconserved elements" remain unaltered for the next generations in these species?

To solve this riddle, genomicists hypothesized that the protein encoding genes are less prone to mutation as it is supposed to alter the profile/configuration of the protein, which might have deleterious effect on the offsprings to kill them or the mutant gene might not pass through the germ cells to the next generation due to premature death before reaching the reproductive age. Initially, the contemporary genomicists contemplated that natural selection acts like an antagonistic measure which stalls mutation in the "Ultraconserved Regions" (Bejerano et al, 2004). So, the scientists tried to extrapolate further that the "Ultraconserved Regions" may not be responsible for encoding proteins but those segments of the DNA seems to act as mutation blocking performance to maintain the archetypic genetic profile/configuration in animals, particularly, in advanced group of vertebrates (Maxmen, 18th January 2018).

In order to decipher the evolutionary function of "ultraconserved elements," Nadav Ahituv, the eminent genomicist at Environmental Genomics and Systems Biology Division, Lawrence Berkley National Laboratory, USA, and his colleagues experimented by removing the 4 non-coding, ultraconserved elements of the 200 base-pair DNA stretches in the mice. In order to maximize the phenotypic trait, Ahituv and his team members (2007) deleted all elements that could act as enhancers in this transgenic assay, so that the near genes exhibiting marked phenotype in both scenarios were either completely activated in the mouse or altered expression due to other genomic factors. At the end of their experiments, the scientists observed that the four lines of mice DNA (having no ultraconserved elements) were found to be viable and fertile as there was no substantial phenotypic abnormalities observed, in terms of growth, life-span, metabolism, pathology etc. among their offsprings but it was expected that in the absence of "ultraconserved elements" those mice should not be viable or fertile (Ahituv et al, 2007). Hence, the earlier observation and hypothesis of Bejerano et al (2004), was found to be not a correct interpretation of identical sequence of non-coding ultraconserved elements of DNA contained by the genomes of a diverse array of vertebrates (Maxmen, 18th January 2018).

Recently, in the year of 2017, Diane Dickel, the contemporary genomicist at Environmental genomics and Systems Biology Division, Lawrence Berkley National Laboratory, USA, along with her colleagues, in an article titled, *"Ultraconserved enhancers are required for normal development,"* observed that deletion of "ultraconserved elements" (the overlapping stretches of DNA that contain genes playing important role in brain development) by usage of precise gene-editing tool called "CRISPR-Cas9" has no impact on viability and reproductive fertility. However, anatomical examination of the brain tissues of the offsprings revealed some brain developmental issues (Dickel et al, 2018). The lack development of brain tissues was observed in the fore-brain of the new-born mice (edited with ultraconserved elements) leading to epilepsy and Alzheimer's disease in matured offsprings (Maxmen, 18th January 2018). Hence, the serious concern of such implications, like cognitive defect, could be transferred to the next generations in the wild population of any higher group of mammalian species if there is no biological safety valve. However it was presumed by the genome biologists that the extent of mutation in the non-coding stretches of DNA or Dark DNA was supposed to be less likely. Even the "Ultraconserved elements" would go through mutation by chance, the reproductive potential of the individuals (with mutated gene of the Dark DNA) seem to fail produce offsprings with edited Ultraconserved elements. The contemporary genomicists thought that Ultraconserved elements might not directly participate in protein encoding but it would interfere and regulate the gene expression of protein encoding genes indirectly (Maxmen, 18th January 2018). So, it was extrapolated by the contemporary genome biologists, genomicists, evolutionary biologists and immunologists that such mutation of any unidentified location in the vast stretches of non-coding Dark DNA would yield a number of neurological and autoimmune diseases such as epilepsy, dementia, holoprosencephaly, Alzheimer's disease, Parkinson's disease etc. and the molecular footprints of the therapeutics related to curing of such diseases should be studied under the contemporary principles of epigenetics (Maxmen, 18th January 2018).

It is mentioned earlier that at the end of the Human Genome Project, in 2001, genome biologists revealed that majority of the genes located in the DNA sequences are involved in RNA coding, whereas, the remaining genes contained are responsible for protein encoding. Even some of the erstwhile protein coding genes were found to be non-functional after being in a fully functional state for a period of time which were recognized as "pseudogenes." To unwind the mystery of the functional mode Dark DNA as well as the genes responsible for RNA coding, molecular geneticists and genomicists engaged in the immunological experimental protocol to implicate the mutational effect on gene expressions. The immunologists have observed for a long time that any structural alteration of DNA (in nucleus, or cytoplasm or in mitochondria) due to mutation may lead to autoimmune disease or make the organism more vulnerable to microbial infections by downgrading its immunological profile (O'Neill, 2013). They also observed that like human systems, microbial systems like bacteria, archaea, protozoa etc. go through the genetic mutations to build defense mechanisms and antibiotic resistance subdue the therapeutic strategies. A number of genomicists tried to explore the therapeutic applications of gene in organisms by modifying the gene expression (application of epigenetics or gene therapy) rather than altering the genetic code of its genome.

In the year 2007-2008, Pawan Kumar Dhar, a scientist in Synthetic Genomics, at RIKEN Genomic Sciences Centre, Yakohoma, Japan, dared to study the genes from "Junk DNA" to explore the field of molecular therapeutics and find whether epigenetics could play a cardinal role in curing genetically inherited disease. As this field of specialization stepped in an uncharted territory, it did not have any existing publication that could prove that epigenetics is not much researched upon. So, Pawan K. Dhar engaged a laboratory technician from Myanmar to team up with his colleagues from Singapore to perform research work in the field of bioinformatics (Gogia, 8th January 2018). Pawan K. Dhar and his colleagues strived to explore the expression of genes screened from "Junk DNA." These genes were technically recognized as "Junk genes" and finally he found the expression of "Junk genes" and was able to decipher the role of "Junk genes" in the survival of living cells. In order to explore the therapeutic potential of Junk genes, Pawan K. Dhar and his associates engaged in data mining of non-coding sequences of DNA and built a massive digital inventory of non-coding Dark DNA sequences of diverse group of organisms in the Department of Computational Biology and Bioinformatics, University of Kerala, India, since 2012, and has continued till date in the School of Biotechnology, Jawaharlal Nehru University, New Delhi, India (Gogia, 27th December 2017). Following the footsteps of Bejerano et al (2004); Ahituv et al, (2007); Roessler et al (2012) and Dickel et al (2018), Pawan K. Dhar and his associates at the Jawaharlal Nehru University have developed a molecular therapeutic application by harnessing the hidden treasure trove of gene expression of "junk genes" in the Dark DNA sequences. Though, the glorified history of "Junk Genes" within the "Dark DNA" has yet to be shared by the Pawan K. Dhar and his associates, a personal communication between Dr. Dhar and Eshna Gogia (27th December 2017)—a well-known science communicator and the business developer of Helixworks—revealed that Pawan K. Dhar and his associates have successfully developed anti-cancer, anti-leishmanial and anti-malarial peptides which have successfully passed clinical trial on human systems. These genomicists engaged in the experimental validation of peptides extracted from Junk DNA sequences and the drugs from the experiments have primarily been discovered to cure fatal neuro-degenerative diseases like Alzheimer's disease and Parkinson's disease (Gogia, 27th December 2017). It seems like the earnest endeavor of these contemporary scientists engaged in Junk-DNA-oriented therapeutic explorations by using laboratory-made genes could help to synthesize "tailor-made proteins" or "custom-made enzymes" having immense potential to cure serious ailments like neuro-degenerative, auto-immune, genetically transmitted diseases etc. The work of these new-era molecular biologists, genomicists and synthetic biologists are trying to build a unique framework of scientific studies pertaining to bioinformatics by proposing as well as introducing intellectual property rights. This is done by assigning distinct intellectual property protocol on the synthesized bio-molecule, the ultimate encrypted, holy grail of each lab-made gene extracted from junk sequence of DNA or Dark DNA, which seems to have a hidden side beyond visual perception.

Very recently, Pavan Vedula, a famous genomicist, led a group of scientists at the University of Pennsylvania and at the National Institutes of Health, USA, to study the dynamics of the wonder protein "Actin." They traced its molecular structure and its multi-tasking role ranging from developmental morphogenesis, cell migration to cell homeostasis, muscular contraction etc. (Vedula et al, 2017). Vedula and his associates (2017) noticed that Actin, the intracellular protein. exists in two iso-forms (which are almost indistinguishable, except for the different 4 amino-acid sequences at the tail end): β Actin and γ Actin and these two iso-forms of protein are encoded by different genes. After critical examination of the homologous iso-forms of Actin, and the genes responsible for encoding these two iso-forms, Vedula et al (2017) came to the conclusion that distinct structure and functions of β Actin and γ Actin are defined by nucleotides, rather than their amino acid sequence. This is because the same base in different sequence form two distinct codes and are recognized as silent codes to encode these two iso-forms. By using the gene-editing tool, CRISPR, Vedula and his colleagues edited the β Actin gene of the mice to stall the production of β Actin protein and noticed the disruption of major functions like embryogenesis, cellular migration etc. (Vedula et al, 2021). Furthermore, they observed that suppression of production of β Actin protein by editing β Actin gene would help to activate the γ Actin gene to produce γ Actin protein which might carry out essential functions like cell migration, embryogenesis and make cells more viable for survival in the organism (Vedula et al, 2017). The researchers contemplated that the amino acid sequence of DNA is not responsible for such functional switch-over of translation dynamics from one iso-form to another form of protein. Rather, its mutational change in gene, specifically, the silent substitutions altered by the functional dynamics of protein translation (Vedula et al, 2021) made Vedula and his fellow scientists (2017) express that, *"our data suggest that the essential in vivo function of β actin is provided by the gene sequence independent of the encoded protein isoform....this regulation constitutes a global 'Silent Code' mechanism that controls the functional diversity of protein isoforms."* In 2017, team Vedula further revealed that different iso-forms of Actin could have different ribosome densities and in the cytoplasm, one form of protein isoform could compensate another isoform when ribosome density is the same (Vedula et al, 2021). This elucidates further that in conditions like differential ribosome density, the loss of β Actin would never be compensated by any other iso-form. Such observation out of this experimental outcome made Vedula et al (2017) wonder whether mutation changes at molecular level like silent substitutions are responsible for the changes in the ribosome density in nucleotide sequence which alter the translation dynamics of protein and affects the rate of protein accumulation in general along with altering the regulations of functional diversity of Actins. In their unique studies, Zhang et al (2010) and Vedula et al (2017) defined the functional mode of iso-form of Actins around the cellular periphery as "Zipcode-mediated transport" and expressed their optimism that there might be a number of proteins that is found in different organisms whose translation dynamics might be controlled by the "Silent Code."

In this regard, it should also be mentioned that the earlier works of the famous molecular biologists helped to predict that the genes encoding the iso-forms of "non-muscle actin" protein, diverged around 100 million years ago as a mutation of randomized, "synonymous substitution" and altering base sequence of the code and resulting in the formation of "Silent Code" (Erba et al, 1986; Zhang et al, 2010). Interpolating the findings of Erba et al (1986), Vedula and his fellow scientists further promulgated that the coding sequence of Actin iso-forms sustained a cumulative effect of "additional evolutionary pressure" to make the amino acid sequence sustainable for a long period of time. They further hypothesized that a part of this "evolutionary pressure" has driven the "divergent translation dynamics" in the family of Actin iso-forms and their "divergent functions" (Vedula et al, 2021). Apparently, the word "Junkomics," used by a popular science writer as well as blogger, Eshna Gogia, in his article titled "*Junkomics: Diggin' through genomic 'garbage' for the hidden treasure*" (published on 27th Dec 2017), sounds like jargon but the present efforts of scientists, in quest of hidden gems to either discover the molecular footprints of life or playing a cardinal role in molecular therapeutics to treat and cure genetically transmitted diseases are going on. So, the word "Junkomics" is not jargon, it's a new identity of the so-called garbage or junk pile of the DNA, of the genome, which was dumped aside and ignored by mainstream researchers in genomics. Junkomics inevitably challenges the current and next generation of "Junkomicists" to explore the secrets of life from a new perspective.

References

Adams, M.D., Celniker, S.E., Holt, R.A., Evans, C.A., Gocayne, J.D., Amanatides, P.G., Scherer, S.E., Li, P.W., Hoskins, R.A., Galle, R.F., George, R.A., Lewis, S.E., Richards, S., Ashburner, M., Henderson, S.N., Sutton, G.G., Wortman, J.R., Yandell, M.D., Zhang, Q., Chen, L.X., Brandon, R.C., Rogers, Y,H., Blazej, R.G., Champe, M., Pfeiffer, B.D., Wan, K.H., Doyle, C., Baxter, E.G., Helt, G., Nelson, C.R., Gabor, G.L., Abril, J.F., Agbayani, A., An, H.J., Andrews-Pfannkoch, C., Baldwin, D., Ballew, R.M., Basu, A., Baxendale, J., Bayraktaroglu, L., Beasley, E.M., Beeson, K.Y., Benos, P.V., Berman, B.P., Bhandari, D., Bolshakov, S., Borkova, D., Botchan. M.R., Bouck, J., Brokstein, P., Brottier, P., Burtis, K.C., Busam, D.A., Butler, H., Cadieu, E., Center, A., Chandra, I., Cherry, J.M., Cawley, S., Dahlke, C., Davenport, L.B., Davies, P., de Pablos, B., Delcher, A., Deng, Z., Mays, A.D., Dew, I., Dietz, S.M., Dodson, K., Doup. L.E., Downes, M., Dugan-Rocha, S., Dunkov, B.C., Dunn, P., Durbin, K.J., Evangelista, C.C., Ferraz, C., Ferriera, S., Fleischmann, W., Fosler, C., Gabrielian, A.E., Garg, N.S., Gelbart, W. M., Glasser, K., Glodek, A, Gong, F., Gorrell, J.H., Gu, Z., Guan, P., Harris, M., Harris, N. L., Harvey, D., Heiman, T. J., Hernandez, J. R., Houck, J., Hostin, D., Houston, K. A. Howland, T.J., Wei, M.H., Ibegwam, C., Jalali, M., Kalush, F., Karpen, G.H., Ke, Z., Kennison, J.A., Ketchum, K.A., Kimmel, B.E., Kodira, C.D., Kraft, C., Kravitz, S., Kulp, D., Lai, Z., Lasko, P., Lei, Y., Levitsky, A.A., Li, J., Li, Z., Liang, Y., Lin, X., Liu, X., Mattei, B., McIntosh, T.C., McLeod, M.P., McPherson, D., Merkulov, G., Milshina, N.V., Mobarry, C., Morris, J., Moshrefi, A., Mount, S.M., Moy, M., Murphy, B., Murphy, L., Muzny, D.M., Nelson, D.L., Nelson, D. R., Nelson, K. A., Nixon, K., Nusskern, D. R., Pacleb, J.M., Palazzolo, M., Pittman, G.S., Pan, S., Pollard, J., Puri, V., Reese, M.G., Reinert, K., Remington, K., Saunders, R.D., Scheeler, F., Shen, H., Shue, B.C., Sidén- Kiamos, I., Simpson, M., Skupskim, M.P., Smith, T., Spier, E., Spradling, A.C., Stapleton, M., Strong, R., Sun, E., Svirskas, R., Tector, C., Turner, R., Venter, E., Wang, A.H., Wang, X., Wang, Z. Y., Wassarman, D.A., Weinstock, G.M., Weissenbach, J., Williams, S.M., Woodage,T., Worley, K.C., Wu, D., Yang, S., Yao, Q.A., Ye, J., Yeh, R.F., Zaveri, J.S., Zhan, M., Zhang, G., Zhao, Q., Zheng, L., Zheng, X.H., Zhong, F.N., Zhong, W., Zhou, X., Zhu, S., Zhu, X., Smith, H.O., Gibbs, R.A., Myers, E.W., Rubin, G.M., and Venter, J.C. (2000). The genome sequence of *Drosophila melanogaster*. Science. 287(5461): 2185-95. doi:10.1126/science. 287. 5461.2185.

Ahituv, N., Zhu, Y., Visel, A., Holt, A., Afzal, V., Pennacchio, L.A., and Rubin, E.M. (2007). Deletion of ultraconserved elements yields viable mice. PLoS Biol. 5(9): e234. doi: 10.1371/journal.pbio.0050234.

Bejerano, G., Pheasant, M., Makunin, I., Stephen, S., Kent, W.J., Mattick, J.S., and Haussler, D. (2004). Ultraconserved elements in the human genome. Science. 304(5675): 1321-5. doi: 10.1126/science.1098119.

Carrol, S.B. (2005). Evolution at two levels: on genes and form. PLoS Biol., 3 (2005): e245.
https://doi.org/10.1371/journal.pbio.0030245

Comings, D. E., and Okada, T. A. (1974). Some aspects of chromosome structure in eukaryotes. Cold Spring Harb Symp Quant Biol. 38:145-53. doi: 10. 1101/sqb.1974.038.01.018.

Dickel, D.E., Ypsilanti, A.R., Pla, R., Zhu, Y., Barozzi, I., Mannion, B.J., Khin, Y.S., Fukuda-Yuzawa, Y., Plajzer-Frick, I., Pickle, C.S., Lee, E.A., Harrington, A.N., Pham, Q.T., Garvin, T.H., Kato, M., Osterwalder, M., Akiyama, J.A., Afzal, V., Rubenstein, J.L.R., Pennacchio, L.A., and Visel, A. (2018). Ultraconserved Enhancers Are Required for Normal Development. Cell. 172(3): 491-499.e15. doi: 10.1016/j.cell.2017.12.017.

Erba, H.P., Gunning, P., and Kedes, L. (1986). Nucleotide sequence of the human gamma cytoskeletal actin mRNA: anomalous evolution of vertebrate non-muscle actin genes. Nucleic Acids Research 14:5275–5294. DOI: https://doi.org/10.1093/nar/14.13.5275

Escobar, J.S., Glémin, S., and Galtier, N. (2011). GC-biased gene conversion impacts ribosomal DNA evolution in vertebrates, angiosperms, and other eukaryotes. Mol Biol Evol. 28(9): 2561-75. doi: 10.1093/molbev/msr079.

Evans, H.H., Evans, T.E. and Littman, S. (1973) Methylation of Parental and Progeny DNA Strands in *Physarum polycephalum*. Journal of Molecular Biology, 74: 563-574.
http://dx.doi.org/10.1016/0022-2836(73)90047-8

Germain, P. L., Ratti, E. and Boem, F. (2014). Junk or functional DNA? ENCODE and the function controversy. Biol Philos 29: 807–831 (2014). https://doi.org/10.1007/s10539-014-9441-3

Gogia, E. (8th January 2018). 'Junk DNA': Mining our genome's dark matter for new disease treatments.
https://geneticliteracyproject.org/2018/01/08/junk- dna-mining-genomes-dark-matter-new-disease-treatments/

Gogia, E. (27th December 2017). Junkomics: Diggin' through genomic 'garbage' for the hidden treasure.
https://nucleicacidmemory.com/junkomics-diggin- through-genomic-garbage-for-the-hidden-treasure-d34bfc0e3926

Hargreaves, A. (25th August 2017). Dark DNA: The phenomenon that could change how we think about evolution.
https://www.business-standard .com/article/current-affairs/dark-dna-the-phenomenon-that-could-change- how-we-think-about-evolution-117082500256_1.html

Hargreaves, A. (2018). The hunt for dark DNA. New Scientist 237(3168): 29-31.
https://doi.org/10.1016/S0262-4079(18)30440-8.

Hargreaves, A.D., Zhou, L., Christensen, J., Marlétaz, F., Liu, S., Li, F., Jansen, P.G., Spiga, E., Hansen, M.T., Pedersen, S.V.H., Biswas, S., Serikawa, K., Fox, B.A., Taylor, W.R., Mulley, J.F., Zhang, G., Heller, R.S., and Holland, P.W.H. (2017). Genome sequence of a diabetes-prone rodent reveals a mutation hotspot around the ParaHox gene cluster. PNAS U S A. 114(29):7677-7682. doi: 10.1073/pnas.1702930114.

Heitz, E. (1928). Das Heterochromatin der Moose. Jahrb Wiss Botanik. 69:762–818.

Hron, T., Pajer, P., Pačes, J., Bartůněk, P., and Elleder, D. (2015). Hidden genes in birds. Genome Biol 16: 164. https://doi.org/10.1186/s13059-015- 0724-z

Kellis, M., Wold, B., Snyder, M.P., Bernstein, B.E., Kundaje, A., Marinov, G.K., Ward, L.D., Birney, E., Crawford, G.E., Dekker, J., Dunham, I., Elnitski, L.L., Farnham, P.J., Feingold, E.A., Gerstein, M., Giddings, M.C., Gilbert, D.M., Gingeras, T.R., Green, E.D., Guigo, R., Hubbard, T., Kent, J., Lieb, J.D., Myers, R.M., Pazin, M.J., Ren, B., Stamatoyannopoulos, J.A., Weng, Z., White, K.P., and Hardison, R.C. (2014). Defining functional DNA elements in the human genome. PNAS U S A. 111(17): 6131-8. doi: 10.1073/pnas.1318948111.

Kieleczawa, J. (2005). Simple modifications of the standard DNA sequencing protocol allow for sequencing through siRNA hairpins and other repeats. J Biomol Tech. 16(3): 220-3.

Kieleczawa, J. (2006). Fundamentals of sequencing of difficult templates--an overview. J Biomol Tech. 17(3): 207-17.

Manuelidis, L. (1990). A view of interphase chromosomes. Science. 250(4987):1533-40. doi: 10.1126/science.2274784.

Masatoshi, N. (2013). Mutation-Driven Evolution. OUP,Oxford. P.256.

Maxam, A.M. and Gilbert, W. (1977). A new method of sequencing DNA. PNAS USA 74(2): 560-564. https://doi.org/10.1073/pnas.74.2.560

Maxmen, A. (18th January 2018). 'Dark Matter' DNA influences brain development. https://www.nature.com/articles/d41586-018-00920-x

Mohlenhoff, B., Romeo, M., Diem, M., and Wood, B.R. (2005). Mie-type scattering and non-Beer-Lambert absorption behavior of human cells in infrared microspectroscopy. Biophys J. 88(5):3635-40. doi: 10.1529/biophysj.104.057950.

Muller, H. J. (1928). The Production of Mutations by X-Rays. PNAS U S A. 14(9): 714-26. doi: 10.1073/pnas.14.9.714.

O'Neill, L. A. (2013). Immunology. Sensing the dark side of DNA. Science. 339(6121): 763-4. doi: 10.1126/science.1234724.

Ris, H., and Kubai, D. F. (1970). Chromosome structure. Annu Rev Genet. 4:263-94. doi: 10.1146/annurev.ge.04.120170.001403.

Roberts, P.A. (1975). In support of the telomere concept. Genetics 80:135–142.

Roessler, J., Ammerpohl, O., Gutwein, J., Hasemeier, B., Anwar, S.L., Kreipe, H., and Lehmann, U. (2012). Quantitative cross-validation and content analysis of the 450k DNA methylation array from Illumina, Inc. BMC Res Notes 5: 210. https://doi.org/10.1186/1756-0500-5-210

Sanger, F., Nicklen, S., and Coulson, A.R. (1977). DNA sequencing with chain-terminating inhibitors. PNAS U S A. 74(12):5463-7. doi: 10.1073/ pnas. 74.12.5463.

Sharma, T., Cheong, N., Sen, P., and Sen, S. (1986). Constitutive heterochromatin and evolutionary divergence of *Mus dunni, M. booduga* and *M. musculus.* Curr Top Microbiol Immunol. 127:35-44. doi: 10.1007/978-3-642-71304-0_4.

The ENCODE Project Consortium. (2012). An integrated encyclopedia of DNA elements in the human genome. Nature 489: 57–74. https://doi.org /10. 1038/nature11247

Swift, H., Adams, B.J., and Larsen, K. (1964). Electron microscope cytochemistry of nucleic acids in *Drosophila* salivary glands and Terahymena. J. R. Microscop. Soc. 83:161–167.

Vedula, P., Kurosaka, S., Leu, N.A., Wolf, Y.I., Shabalina, S.A., Wang, J., Sterling, S., Dong, D.W. and Kashina, A. (2017). Diverse functions of homologous actin isoforms are defined by their nucleotide, rather than their amino acid sequence. Elife. 6: e31661. doi: 10.7554/eLife.31661.

Vedula, P., Kurosaka, S., MacTaggart, B., Ni, Q., Papoian, G., Jiang, Y., Dong, D.W., and Kashina, A. (2021). Different translation dynamics of β- and γ-actin regulates cell migration. Elife. 10: e68712. doi: 10.7554/eLife .68712.

White, M.A. Myers, C.A., Corbo, J.C., and Cohen , B.A. (2013). Massively parallel in vivo enhancer assay reveals that highly local features determine the cis-regulatory function of ChIP-seq peaks. PNAS USA 110 (29):11952-11957; DOI: 10.1073/pnas.1307449110

Yamashita, J., and Kamen, M. D. (1968). Observations on the nature of pulse-labeled RNA's from photosynthetically or heterotrophically grown *Rhodospirillum rubrum.* Biochim. Biophys. Acta 161:162–169.

Zhang, F., Saha, S., Shabalina, S.A., and Kashina, A. (2010). Differential arginylation of actin isoforms is regulated by coding sequence-dependent degradation. Science. 329(5998):1534-7. doi: 10.1126/science.1191701.

Chapter 8

Dark DNA: In Quest of Evolutionary Footprints of Life - I

Dark DNA is thought to contain large regions of mutational hotspots. Such hotspots undergo high mutation rates. Since evolutionists view mutations as the genetic engine driving dramatic transformations, dark DNA could be a source of rapid and extensive evolutionary change. However, mutations have never been shown to accomplish the types of transformations required by universal common descent. Thus, dark DNA is not a special nursery for evolutionary change.

---- Kevin Anderson

A take-home message for biologists, evolutionists, researchers, academics, students, directly or indirectly related to the community of science is that nucleic acids, specifically the DNA sequence is the holy grail of molecular genetics and genomics which could flawlessly define the molecular footprints of life, unfurling all its mysteries and decrypting the genetic fate of any organism. In the field of science, there is hardly any "truth" as each discovery brings about a shift in paradigm. Thus, the identity of scientific perception has a noticeably short-span of life and is dynamic in nature.

In order to grasp the ideas of evolutionary function and its relation to adaptation and natural selection—the two cardinal pillars of evolution—we need to review the case studies of organisms that inherited adaptive evolution to meet survival challenges. Let's go through a few examples of survival strategies that some organisms adapted in unnatural environmental conditions. One such example is the catfish or the *Typhlichthys subterraneus*. It is popularly known as the Southern cave fish endemic to the southern part of North America namely Alabama, Missouri, Arkansas, Oklahoma and Kentucky. They are found in underground water channels (Poulson, 1963; Jones and Taber, 1985). As these cave fish live in absolute darkness, their eyes have degenerated under special adaptive conditions. The temperature in the caves varies from 10 to 15 degree Celsius in different season and there is always scarcity of food. All these microclimatic conditions have made the blind catfish look like white, pigment less fish which is an adaptation to cut-down excess metabolic expenditure (Jones and Taber, 1985). Another example of dormancy where a number of animals adopt unique measures to survive challenges are the wood frogs (*Rana sylvatica*). They are found in the Arctic eco-climatic regions of North America and Northern Europe and are known to produce anti-freezing protein (AFPs) (Costanzo and Lee, 2005). When the temperature drops below -30 to -40 degree Celsius, the anti-freezing protein prevents the water fluid or blood from crystallization which could have depleted the entire frog population. During the production of the anti-freezing protein, regular functions such as breathing, bleeding, movement, eating etc. and physical and physiological actions including normal metabolism is put on hold. Thus, the wood-frogs in temperate and semi-arctic forest ecosystems are found in dormant state from the beginning of winter till spring (Costanzo and Lee, 2005). The two examples made the scientists, particularly, the evolutionists, happy as they understood and further defined the role of adaptation and natural selection which acts as the functional force of biological evolution. However, it did not fully satisfy those who strived to look past the veil which hid the molecular footprints of life to understand the function of biological evolution of life in a broader aspect.

Naturally, the ideas, interpretations and courses of action of the contemporary evolutionary biologists were biased towards the findings and interpretations of molecular data such as decrypting the genome of a particular species and its comparison to the genome of a related species. Unfortunately, the conventional revelations of nucleic acids (specifically DNAs) would not be enough to debrief the molecular footprints of evolution. After going through a cascade of shifts in paradigm, from Mendelian principle of genes, the principle of genes in the nucleoids, the non-coding sequences of DNA, Junk DNA to Dark DNA, the scientists came to understand and let others narrate the functional aspects of biological evolution. The resemblance of Dark DNA in the biological world to Dark Matter in our universe is another attempt to de-mystifying the identity of a large chunk of genome (apparently non-functional) devoid of encoding any functional proteins (Perry, 29th August 2017). The diagrammatic sketch of Dark DNA has been presented in fig. 8.1.

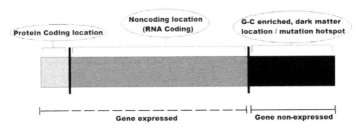

Fig. 8.1. The diagrammatic sketch of Dark DNA.
(Modified from Gogia, 2018; https://geneticliteracyproject.org)

According to the basic principles of Dark DNA defined by genomicists, the phenotypic expression of a protein or enzyme (a trace of it or in the modified form) might be witnessed in an organism, in which, encoding gene of the protein is not found at all. However, subsequent molecular investigations found the gene to be there after screening for their presence in the genomes of genetically close-related organisms (which is devoid of that gene). So, it indirectly ascertained that the undetectable gene must be there in the genome of the aforesaid organism as an integral constituent of Dark DNA.

In order to present an analogical resemblance between Dark Matter of the universe and the Dark DNA we must scrutinize the elusive nature of both the entities. Astronomers estimate that about 25% presence of Dark Matter in this universe, substantiated by advanced physical experiments, is invisible. They have further hypothesized that erstwhile existence of matter and energy must be more than its current level of existence (Anderson, 1st November 2018). Though detection of Dark Matter and Dark Energy is found to be a cumbersome perception but the combined application of theoretical reasoning in Physics and the level of technological advancement not only helped to prove its existence but also to explain the cosmological model related to the formation of universe. The use of "dark substance' in the biological world is a stretch. With reference to the dark biological substances, the term "Microbial dark matter" is often mentioned. This phenomenon has hardly been observed in laboratory conditions rather than in certain environmental conditions, but the existence of such a state is proven by DNA or RNA. Most of the time, the term "biological dark matter" is attributed to the part of the DNA whose function is not yet defined (Anderson, 1st November 2018).

In contemporary research, a group of genomicists led by Adam

Hargreaves (25th August 2017), engaged in deciphering the identity of Dark DNA in terms of its structure and function in the sand rat (*Psammomys obesus*)—a hot desert dwelling gerbil found in North Africa and Middle-East Asia. The scientists involved in studying the gene activity discovered that the gene contained non-functional and invisible stretches of DNA. After completion of DNA sequencing of the sand rat genome, they noticed that around 88 genes were missing from the "undetected" or difficult to detect stretches of DNA of this rodent but was present in the chromosomal sequence data of its phylogenetic allies ((Hargreaves et al, 2017). They referred to the "undetected" and "non-coding' stretches of DNA as Dark DNA and which was also found in the genome of chicken and gerbil. Thus, it needed to be ascertained that this stretch of genome of sand rat was not missing, rather it was invisible due to failure of conventional DNA sequencing protocols (Hargreaves, 2018). Critical review revealed that a number of genes contained in the Dark DNA region appeared to be in different sequence in comparison to their counterparts in other species (Anderson, 1st November 2018). The genomicists, particularly those engaged in studying the susceptibility of the gerbil to type-2 diabetes and the role of insulin production genes such as *Pdx1* gene, needed to isolated the genome of the phylogenetically-related mammal species of the gerbil, as *Pdx1* gene was found to be responsible for regulating the secretion of insulin hormone in other organisms. After repeated examinations, they failed to trace the existence of *Pdx1* gene in the genome of the gerbil along with 87 genes in its vicinity (J-Mac, 16th March 2018).

The most confusing fact for the genomicists was the absence of *Pdx1* gene in the genome of sand rat. It is most unlikely as the absence of this essential gene would mean that gerbil would not survive under normal condition. Hence, the survival of the sand rat, provided a hint to Hargreaves et al (2017) that the insulin encoding *Pdx1* gene was not missing, rather, it was hidden somewhere in the genome of gerbil. The biochemical investigations of a different part of the tissue from a sand rat in the next level ascertained that the biochemical (proteins and its derivatives) encoded by the *Pdx1* gene was present and the production of those proteins and protein derivatives would not be possible without being regulated by the presence of *Pdx1* gene (Perry, 29th August 2017; J-Mac, 16th March 2018). This confirmed that an essential gene like *Pdx1* is not missing or absent, rather, it exists in the genome of gerbil in hidden condition.

The critical review of the genomicists, led by Hargreaves on the genomic profile of the sand rat and the comparative studies of the sand rat with its phylogenetically close relative rodent species revealed that the genome of sand-rat has gone through more mutations than its phylogenetic allies (Perry, 29th August 2017). Furthermore, scientists hypothesized that the essential genes—those that ensured the survival of this species—was located on the genome and are called "hotspots" (Perry, 29th August 2017). We are familiar with the base-pairings of the nitrogen bases in deoxy-ribonucleotides where purine base Adenine (A) pair with pyrimidine base Thymine (T) and purine base Guanine (G) pair with pyrimidine base Cytosine (C). In terms of DNA sequencing, which helps determining the sequence of nucleotide (sequential array of A:T, G:C base-pairings) in DNA stretches, biotechnologists faced problem sequencing the GC-rich genome. Based on their research on the genetic make-up of the sand rat genome, particularly the identity and location of the "missing genes." Hargreaves and his associates (2017) observed that preponderance of GC in the Dark DNA and the high content of GC base pairs made it difficult to sequence DNA in the Dark DNA stretch of the concerned organism (Anderson, 1st November 2018). Hargreaves and his colleagues identified the GC-rich locations of the gerbil genome as the ideal abode for those missing genes. They further hypothesized that all those "missing genes" are difficult to be traced as these locations act as mutation hotspots that go through a series of random mutations leading to the loss of its erstwhile identity. Those essential genes play a cardinal role in the survival of the concerned species but appear in a derived form which made it difficult to recognize its existence and created perception of "missing genes" (Perry, 29th August 2017).

However, a comparative account of the genetic make-up, particularly the GC-rich location of genome of the gerbil with its close phylogenetic ally, has given an impression that random mutation effect such as "biased gene conversion" (i.e. base-pair mismatch during post-recombination phase of DNA) leads to the formation of GC-rich location in the sand rat genome (Moran, 18th March 2018). Moreover, the resemblance of the GC-rich location in identical places between the genome of gerbil and its allies also helped scientists to theorize a new evolutionary interpretation of the "Dark DNA." They hypothesized that a shift in preponderance of GC content occurred long before the gerbil and its phylogenetic ally emerged from their last common ancestor (Moran, 18th March 2018).

In order to define the potential role of Dark DNA, it is necessary to understand whether it is restricted to the genome of a few organism or ubiquitous and present in all higher groups of eukaryotic animals. Though, the term "Dark DNA" is not a colloquial term in the field of biology but it is not an unprecedented word either as it existed in avian species (Perry, 29th August 2017; Moran, 18th March 2018). A group of genomicists under Peter V. Lovell (2014) carried out a comparative genomic investigation on 60 avian genomes and found that around 274 protein encoding genes were "missing" from the sequenced bird genomes but were present in the genomes of most vertebrates. They were specifically observed in the "conserved synthetic cluster" of non-avian dinosaurs and human species. A large number of these missing genes are found to be linked to the lethality of smaller vertebrates like mice and genetic disorders that are also witnessed among human species. From an evolutionary perspective, the "missing genes" were found in the location of the genome of non-avian vertebrates where chromosomal rearrangements and the disappearance of these "missing" genes from the avian genome took place during phylogenetic split of two distinct evolutionary lineages of dinosaurs (which belonged to reptiles) and birds from the mother-stock of "Crocodilians" during the end of Jurassic era around 160-150 million years ago (Lovell et al, 2014).

The presence of highly compact genomes in comparison to other amniotes was initially dubbed as an instance of volant adaptation to cope with excessive rate of oxidative metabolism of avian species that supported their evolution (Hughes et al, 2008). Identical genomes in non-avian dinosaurs indicated that emergence of compact genomes supposedly emerged much before the divergence of flying birds from the crocodilian stock (Organ et al, 2007). According to contemporary analysis of the research on comparative genomics on vertebrates which comprised mammals (human species, rodents etc.), reptiles (dinosaurs, crocodiles etc.) and birds, (chickens, finches etc.) genomicists hypothesized that reduction of genome size in the organisms supposedly expedited due to the loss of non-coding sequences of DNA. This was supported by unique configuration of avian genomes in comparison to larger genomes of mammals which contained few pseudogenes, shorter introns and "less repetitive DNA" (Hillier et al, 2004; Organ et al, 2007). However, the GC-rich DNA or Dark DNA is found in these missing genes and are prevalent in few groups of organisms like bacteria, reptiles, birds and mammals (Perry, 29th August 2017).

In order hypothetically explain the difficulty of DNA sequencing techniques of the GC-rich location, the Dark DNA location of the genome was tendered by Hargreaves in 2018. They put forth the explanation that Dark DNA goes through a barrage of random mutations that alter its sequence with reference to its analogues genes found in the genomes of their close phylogenetic allies. The aftermath of these cascade of mutations leads to an increasing of GC content in the Dark DNA location which problematized the sequencing for genomicists in GC-rich location of DNA. Such interpretation advocated by Hargreaves's team (2017) tried to present "Dark DNA" in its mutational hotspot where the rate of mutation was higher than the regular cycle of mutation in normal environmental conditions (Anderson, 1st November 2018). However, their ultimate goal pertains to the evolutionary role of Dark DNA which is not necessarily related to tracing the identity of missing genes or discovering their functional role but to establish the functional identity of the "hotspots." Hargreaves and his colleagues (2017) came up with a unique hypothesis which proposes "batch mutation of genes" rather than "mutation of individual gene" as it suggests that the GC-rich mutational hotspots in the genome (containing the dark DNA) act as the ideal "vacuous" mutational abode where a number of genes interact with each other to synthesize essential proteins for the survival of the organism. Instead of mutation of individual gene it's the cumulative mutation of a number of genes in the Dark DNA which, presumably, has driven evolution (Perry, 29th August 2017). In this regard, it should also be mentioned that a group of contemporary evolutionists considered mutations as a huge potential that coordinated changes of genetic configurations in an organism. Being a storehouse of mutational hotspots, Dark DNA is supposed to have the potential to coordinate evolutionary changes by means of rapid and wide extent of mutational changes (Anderson, 1st November 2018). When and if mutation offers a greater chance at altering advantageous physical traits, their cumulative result would render an emergence of "greater variety of traits" (Hargreaves, 2018). Greater variety of traits in comparison to a single or few traits in a population must be considered an evolutionary edge for the concerned organism which would have better chance at natural selection (Anderson, 1st November 2018). Theoretically, such explanation complies with the evolutionary principles of Darwin.

Now, a number of evolutionary biologists and evolutionists have tried to figure out what made Hargreaves and his fellow scientists (2017) take a firm stand in defining the mutational hotspot as the key which regulates evolutionary progress. The main concern was whether the hypothesis had any rational or theoretical background or was it an instance of radical interpretation of a biased observation. In 1930, Sir Ronald A. Fisher, the famous British evolutionary biologist, promulgated Fisher's (1930) "Fundamental Theorem of Natural Selection" in his book titled "*The Genetical Theory of Natural Selection*" and defined this principle under the theoretical elucidation of "Law of Entropy" in fundamental Physics. A number of evolutionary biologists, evolutionists and phylogeneticists consider "Fisher's Theorem" as a statistical interpretation of Darwin's theory of natural selection. According to Fisher's theorem, to maintain an uninterrupted pool of genetic variation in a given population, an uninterrupted influx of mutations needs to be the main concern for maintaining sustainability of any evolutionary dynamics (Lourenco et al, 2016). In his theorem, Fisher (1930) further theorised that constant influx of mutations which helped to generate variations in a given population helped to enhance the level of fitness of the concerned organism. The main drawback of this theorem is that Fisher never calculated the quantitative assessment of mutational dynamics (the rate of mutation of a given population at a given time) by considering issues like tallying the harmful and helpful mutational variations (Lourenco et al, 2016). Instead, Fischer considered that occurrence of helpful mutations were found to be always more prevalent than harmful mutational variations. He further contended that preponderant influx of beneficial mutations would weed out the harmful effects of deleterious mutations which helped the population maintain an equilibrium (Lourenco et al, 2016).

In reality, Fisher's (1930) "Fundamental Theorem of Natural Selection" was in a state of "Catch 22" as it was neither proved to be absolutely invalid nor is it accepted as a valid theory in the field of molecular genetics and genomics. According to the scientific community, high rate of mutation tends to bring down the rate of fitness of a small-sized population (Basener and Sanford, 2018). Though a constant low-level impact of negative fitness would always be there, occurrence of higher rate of mutations in a bigger population would lead to beneficial mutations in the population and ultimately be helpful in increasing fitness of the concerned organism (Hargreaves, 2018). Though DNA is the holy grail of the molecular footprint of most organisms, the oxidative stress is a common feature in our environment. On top of this, factors like anthropogenic pollution, over-exploitation of any organism, its habitat disturbance etc. would make the situation more difficult. As long as we consider the sustainability of the organism and the dynamic equilibrium of molecular evolution as a whole, the resulting shrinkage of gene pool, genetic erosion and abrupt decreasing of population sizes of most organisms, mainly animals, is common (Sanford, 2014). In this era of "Anthropocene," the likelihood of the beneficial mutations are lesser in respect to deleterious mutations. Hence, Ronald Fisher's interpretation of equilibrium of deleterious mutations by influx of healthy mutations holds true in the present perspective (Sanford et al, 2018).

In order to redefine the occurrence of mutation and its role in driving the evolution of an organism, Hargreaves and his associates (Moran, 18th March 2018) tried to portray the functional aspects of Dark DNA in a coherent manner. According to them, the occurrence of mutation is a continuous but random process and natural selection determines whether the effect of mutation is harmful enough to be instantly discarded or beneficial to be carried forward to the next generation which is determined in terms of the "reproductive success" of the concerned organism (Moran, 18th March 2018). In the article titled "Dark DNA: The missing matter at the heart of nature," published in the New Scientist, Hargreaves and his fellow genomicists, speculated that due to the high pace of mutational changes in the Dark DNA, natural selection might not be able to combat the deleterious mutations accumulated (as mutational changes might have a mixed outcome which could be helpful or harmful) in the organism which would jeopardize its existence (Bromham, 2009). As a matter of fact, helpful or harmful mutational variation depends upon the ambience the organism belongs to; so the deleterious or harmful mutational changes could be proven helpful if the dynamics of its environment change in the future. Hence, the concept of harmful and helpful mutation is a perspective which changes with the passage of time and space. As theory does not play a supportive role in this case, hence, referring to deleterious mutations as helpful in the future seems to be speculation which does not have enough validity as we'd never know whether changes in the environment would be better or worse for the concerned organism. So if we need to make a judgement from the perspective of the present scenario, occurrence of deleterious mutations would impose negative effects on the organism and the occurrence of recurrent mutations would accumulate harmful phenotypic effects and pose a serious threat to the health of the entire population of that organism (Anderson, 1st November 2018). Apparently, more insight and refinement of the functional role of GC-rich stretches of DNA is required to legitimise the "Dark DNA driven evolutionary hypothesis" as a feasible interpretation given by Hargreaves and his fellow genomicists (2017). As it seems that contemporary interpretation would not efficiently define the heuristic model of Dark DNA as the main engine would appear to expedite extinction instead of driving the evolutionary progress. We cannot jump to a conclusion by refuting the contemporary concepts of Dark DNA by referring to the present in the backdrop of our conventional ideas either, as it requires extensive investigation to find the presence and extent of diverse array of Dark DNA in different groups of organisms, particularly, the animals and more reviews on the functions of mutational hotspots (i.e. GC-rich non-coding location) of Dark DNA in the future.

Now, in terms of considering GC-rich area of the DNA as a "hotspot" in the genome of an organism, Hargreaves et al (2017) stipulated that the genes in the "hotspot" would have higher chance of mutation. When the mode of mutation was found to taken place by "interceded genes," or regarded as "batch mutations" of genes rather than "individual mutation," the post-mutational effect appeared like a state of "atomic chain reactions" in Physics. Consequently, this was a key driving force to steer evolution in a certain direction (Moran, 18th March 2018). Hence, the logical explanation was that mutational hotspots of the Dark DNA should be recognized as a cardinal regulator of evolutionary transformations. Due to the active role in regulating evolutionary transformations, the mutational potential of the Dark DNA is considered as "an under-appreciated mechanism" by a group of contemporary evolutionists (Wright et al, 2006).

So, Hargreaves et al (2017) have tried to uphold the principles of mutation as one of the key regulating factor along with the role of natural selection which drives the evolution of an organism. However, the unique ideas of Hargreaves et al (2017) appeared to be a conjecture as the existence of Dark DNA was only reported from a few group of organisms (like bacteria, birds, mammals, reptiles etc.) and its absence in most organisms would not validate the clarification of the hypothetical role of Dark DNA or hotspots of Dark DNA to steer evolutionary progress. Apparently, the Dark DNA-driven evolution in the biological world seems to be conjecture due to lack of enough evidence which help substantiate the hypothesis of

Hargreaves (2018). However, it seems that GC-rich hotspots of Dark DNA functionally resembles an air jet pump in a water tank which creates air bubbles in the water and the water molecule in the vicinity of this air jet turn around faster than the distant water molecule. So, it might be a matter of chance that gathering a number of genes in the vicinity of mutation hotspot would trigger evolution where an organism might emerge with an advanced formation and function to keep the wheel of evolution turning.

However, all hypotheses require substantial evidence for it to be accepted as a scientific principle. Refuting the hypothesis of Hargreaves et al (2017), his critics challenged the idea that "Dark DNA" is the site of evolutionary cradle and argued that the impact of mutation was found to be detrimental to the survival of organisms rather than contributing to evolutionary advantage that emerged in a new form (Anderson, 1st November 2018). Furthermore the critics of propositions of Hargreaves et al (2017) argued that if we consider Dark DNA as mutational hotspot for genes, it is the responsibility of the proponents of the hypothesis to prove that the major evolutionary changes like origin of vertebrates from invertebrates or evolution of mammals emerging after split of the common stock of amphibians, reptiles, avis etc. has gone through a shift in paradigm where Dark DNA could contribute to such emergence of distinct group of organisms from the common vertebrates mother stock (Anderson, 1st November 2018). Though theoretical genomics tried to patronize the potential of mutations to achieve the primary target of transformations carried out by the "universal common descent[11]," orthodox denialists of the Dark DNA hypothesis would not accept that the existence of Dark DNA not only acts as biological reactor of mutations that expedite physical changes, but it might also have given birth to advanced genetic structures and functions which emerged as a cradle of evolutionary changes in the future (Anderson, 1st November 2018). In order to define mutations as the prime emerging regulator of "universal common descent" its power needs to substantiate that it could drive major evolutionary transformations in aquatic species to diverge into amphibian and reptilian species and from the stock of reptile species, diverge into flying bird species etc. (Anderson, 1st November 2018).

---

[11] Universal Common Descent: The diverse array of life, diverged from a single, ancestral mother-stock.

Though Ronald Fisher's "Fundamental Theorem of Natural Selection" is still admired by the followers of Neo-Darwinism (Fisher, 1930), contemporary critics of the hypothetical interpretation of mutation-driven evolution critically reviewed the state of population fitness under the influence of random mutations. They came to a conclusion that mutation-induced genetic variations could not offer the required level of genetic changes that could drive "universal common descent" (Anderson, 1st November 2018). Evolutionary biologists further observed that even mutations which were able to define genetic variations could give an explanation of structural and functional alterations of a particular biological organ or a group of organs (like lungs, eyes, flippers etc.) of an organism, but it did not give any clear idea of ontogenesis of an organ or group of organs. It failed to give any clue of structural transformations (either anatomical or morphological) that could reveal the evolutionary footprint of "Universal common descent" (Martins and Posada, 2012; Anderson, 18th August 2016). It's quite surprising to observe that the beneficial mutations which provide selective advantage, have also been found to be degenerative as the mutational benefit of the organisms such as HIV resistance, antibiotic resistance etc. would be achieved at the cost of loosing or inhibiting the erstwhile genetic functions. Such analogical erasure of pre-existing memories of genetic function could hardly create any evolutionary progress in the "universal common descent" for new groups of organisms (Martins and Posada, 2012; Anderson, 18th August 2016).

However, in the history of any specialization, whether scientific innovations or a successful execution of social revolution, nothing is beyond controversy. The consideration of GC-rich stretches of DNA, as well as the non-coding Dark DNA as the hotspot of mutation of genes and its ultimate role in evolution by Hargreaves et al (2017), is not above such controversies. Still, Hargreaves and his colleagues consider that their experimental model of sand-rat has traversed through the evolutionary trail triggered by the mutational "hotspot," the Dark DNA (Perry, 29th August 2017). Logic follows that if mutations don't have any driving potential for common descents from which the derived evolutionary groups like amphibians and reptiles diverged from the fishes in course of evolution, then consideration of the functions of mutational hotspots as "under-appreciated mechanism" should be reviewed and re-defined (Anderson, 1st November 2018). In the present situation, the existence of Dark DNA as non-coding stretches of the genomes in some advanced group of organisms must not be ignored as scientists have observed that Dark DNA played a cardinal role in a number of genetic changes in the genetic profile of the organism. It supports adaptation which is an important part of evolution but the integration of major issues such as—

i.   whether Dark DNA acts as a cradle for new genes or not,

ii. whether it would have regulatory potential to drive molecular evolution by enhancing the level of fitness of the population or not

iii. whether it could control the entire genetic system for defining the evolutionary changes as per Darwinian principle or not

— needs to be settled further (Anderson, 1st November 2018; Moran, 18th March 2018). It seems that the Dark DNA has gone through conflict of ideas, opinions, interpretation of evidence between protagonists and antagonists of the findings and observations of the scientists which helped to reveal the true identity of Dark DNA in future and help the community of science to understand its molecular function. Still, protagonists such as Adam Hargreaves (25th August 2017, 2018), Dan Graur (1985, 2017), Alexander Palazzo and T. Ryan Gregory (2014), P. Z. Myers (2008), Francis S. Collins (2007), Sydney Brenner (8th December 2002), Joseph Felsenstein (2001, 2003), Kenneth R. Miller (1994), W. Ford Doolitile and Carmen Sapienza (1980), Leslie E. Orgel and Francis H. Crick (1980), Richard Dawkins (1976, 1998), Susumu Ohno (1972), David E. Comigs and Evangelita Avelino (1972), Motoo Kimura (1968) etc. consider that the perception of "Dark DNA-driven evolution" is right and in this era of "Anthropocene," the functional involvement of micro-management of Earth's ecosystem by human species would lead to expediting inconceivable level of climate change which would tilt the dynamics of evolutionary changes for a number of species while expecting the prevalence of Dark DNA among more groups of organism in future (Perry, 29th August 2017).

References

Anderson, K. (18th August 2016). Just how random are mutations?
https://answersingenesis.org/genetics/mutations/just-how-random-are-mutations/

Anderson, K. (1st August 2018). Is dark DNA evoultion's secret weapon?
https://answersingenesis.org/genetics/mutations/dark-dna-evolutions-secret-weapon/

Basener, W.F., and Sanford, J.C. (2018). The fundamental theorem of natural selection with mutations. J. Math. Biol., 76 (7): 1589-1622.

Brenner, S. (8th December 2002). Nobel Lecture: Nature's gift to science. Pp.274-282.
https://www.nobelprize.org/uploads/2018/06/brenner- lecture.pdf

Bromham, L. (2009). Why do species vary in their rate of molecular evolution? Biol Lett. 5(3): 401-4. doi: 10.1098/rsbl.2009.0136.

Collins, F.S. (2007). The language of God: A scientist presents evidence for belief (Repr. edn.). Free Press. P. 320.

Comings, D., and Avelino, E. (1972). DNA Loss during Robertsonian Fusion in Studies of the Tobacco Mouse. Nature New Biology 237:199 (1972). https://doi.org/10.1038/newbio237199a0

Costanzo, J. P. and Lee, R. E. (2005). Cryoprotection by urea in a terrestrial hibernating frog, J. Exp. Biol. 208: 4089-4089.

Dawkins, R. (1976). The Selfish Gene. OUP, New York. P. 224.

Dawkins, R. (1998). Unweaving the rainbow: science, delusion and the appetite for wonder. Mariner Books. P. 352.

Doolittle, W.F., and Sapienza, C. (1980). Selfish genes, the phenotype paradigm and genome evolution. Nature. 284(5757): 601-3. doi: 10.1038/284601a0.

Felsenstein, J. (2001). The troubled growth of statistical phylogenetics. Syst Biol. 50(4): 465-467.

Felsenstein, J. (2003). Inferring Phylogenies (3rd Edn.). OUP USA. p.580.

Fisher, R. A. (1930). The genetical theory of natural selection. Oxford: Clarendon Press, New York: Dover.

Graur, D. (1985). Amino acid composition and the evolutionary rates of protein-coding genes. J Mol Evol. 22(1):53-62. doi: 10.1007/BF02105805.

Graur, D. (2017). An Upper Limit on the Functional Fraction of the Human Genome. Genome Biol Evol. 9(7): 1880-1885. doi: 10.1093/gbe/evx121. Erratum in: Genome Biol Evol. 2019 Nov 1;11(11):3158.

Hargreaves, A. (25th August 2017). Dark DNA: The phenomenon that could change how we think about evolution.
https://www.business-standard .com/article/current-affairs/dark-dna-the-phenomenon-that-could-change- how-we-think-about-evolution-117082500256_1.html

Hargreaves, A. (2018). The hunt for dark DNA. New Scientist 237(3168): 29-31. https://doi.org/10.1016/S0262-4079(18)30440-8.

Hargreaves, A.D., Zhou, L., Christensen, J., Marlétaz, F., Liu, S., Li, F., Jansen, P.G., Spiga, E., Hansen, M.T., Pedersen, S.V.H., Biswas, S., Serikawa, K., Fox, B.A., Taylor, W.R., Mulley, J.F., Zhang, G., Heller, R.S., and Holland, P.W.H. (2017). Genome sequence of a diabetes-prone rodent reveals a mutation hotspot around the ParaHox gene cluster. PNAS U S A. 114(29):7677-7682. doi: 10.1073/pnas.1702930114.

Hughes, W.O., Oldroyd, B.P., Beekman, M., and Ratnieks, F.L. (2008). Ancestral monogamy shows kin selection is key to the evolution of eusociality. Science. 320(5880):1213-6. doi: 10.1126/science.1156108.

J-Mac, (16th March 2018). Dark DNA and the new mechanism of Evolution? http://theskepticalzone.com/wp/dark-dna-and-the-new-mechanism-of-evolution/

Jones, S. R., and Taber, C.A. (1985). A range revision for western populations of the southern cavefish *Typhlichthys subterraneus* (Amblyopsidae). American Midland Naturalist 113:413-415.

Kimura, M. (1968). Evolutionary Rate at the Molecular Level. *Nature* 217: 624–626. https://doi.org/10.1038/217624a0

Lourenço, N., Pereira, F.B. and Costa, E. (2016). Unveiling the properties of structured grammatical evolution. Genet Program EvolvableMach 17: 251–289. https://doi.org/10.1007/s10710-015-9262-4

Lovell, P.V., Wirthlin, M., Wilhelm, L., Minx, P., Lazar, N.H., Carbone, L., Warren, W.C., Mello, C.V. (2014). Conserved syntenic clusters of protein coding genes are missing in birds. Genome Biol. 15(12):565. doi: 10. 1186/s13059-014-0565-1.

Martins, L.D., and Posada, D. (2012). Proving universal common ancestry with similar sequences. Trends Evol Biol. 4(1):e5. doi: 10.4081/eb.2012.e5.

Miller, K.R. (1994). Life's Grand Design. Technology Review 97(2): 24-32.

Moran, L.A. (18th March 2018). What is "Dark DNA"? https://sandwalk.blogspot.com/2018/03/what-is-dark-dna.html

Myers, P. (2008). Hox genes in development: The Hox code. Nature Education 1(1):2. http://scienceblogs.com/pharyngula/2007/09/the_hox_c ode.php

Ohno, S. (1972). So much "junk" DNA in our genome. Brookhaven Symp Biol. 23: 366-70.

Organ, C.L., Shedlock, A.M., Meade, A., Pagel, M., and Edwards, S.V. (2007). Origin of avian genome size and structure in non-avian dinosaurs. Nature. 446(7132): 180-4. doi: 10.1038/nature05621.

Orgel, L.E., and Crick, F. H. (1980). Selfish DNA: the ultimate parasite. *Nature*. 284: 604-7. PMID 7366731 DOI: 10.1038/284604A0

Palazzo, A. F, and Gregory, T. R. (2014). The Case for Junk DNA. PLoS Genet 10(5): e1004351. https://doi.org/10.1371/journal.pgen.1004351

Perry, P. (29th August 2017). "Dark DNA" is changing the way we see the evolution.
https://bigthink.com/dark-dna-is-changing-the-way-we-see-evolution

Poulson, T.L. (1963). Cave adaptation in Amblyopsid Fishes. The American Midland Naturalist. 70(2): 257-290.

Sanford, J.C. 2014. Genetic Entropy (4th Edn.). FMS Publications, Waterloo, NY. P. 270.

Sanford, J., R. Carter, W. Brewer, J. Baumgardner, B. Potter, and J. Potter. (2018). Adam and Eve, designed diversity, and allele frequencies. In Proceedings of the Eighth International Conference on Creationism, ed. J.H. Whitmore, pp. 200–216. Pittsburgh, Pennsylvania: Creation Science Fellowship.

Wright, W. D., Shah, S. S., and Heyer, W. D. (2018). Homologous recombination and the repair of DNA double-strand breaks. J Biol Chem. 293(27): 10524- 10535. doi: 10.1074/jbc.TM118.000372.

Chapter 9

Dark DNA: In Quest of
Evolutionary Footprints of Life - II

Some years ago I noticed that there are two kinds of rubbish in the world and that most languages have different words to distinguish them. There is the rubbish we keep, which is junk, and the rubbish we throw away, which is garbage. The excess DNA in our genomes is junk, and it is there because it is harmless, as well as being useless, and because the molecular processes generating extra DNA outpace those getting rid of it. Were the extra DNA to become disadvantageous, it would become subject to selection, just as junk that takes up too much space, or is beginning to smell, is instantly converted to garbage by one's wife, that excellent Darwinian instrument.

---- Sydney Brenner

Since the discovery of Mendelian principle of inheritance in the 19th century to the contemporary regime in the 21st century, the progress of genetics has traversed through a long way of hypothetical conflicts. While evolutionary biologists along with population biologists have felt congenial with the apparent "non-functioning" state (neither in terms of gene regulated phenotypic expressions nor promoting evolutionary changes to a certainty) of a large part of genome for more than last five decades, the classical geneticists and their disciples nurtured old ideas that each nucleotide of a genome is functionally active and contribute directly in driving the evolutionary progress. One such orthodox geneticist, Vogel (1964) assessed around 6.7 million protein-encoding genes contained by the human genome, and the result was excessively high in number which was clarified by him as two hypothetical possibilities:

    a. It happens in the genomes of higher group of organisms, having large part of structural genes than protein-encoding genes.

b. Protein-encoding genes in human species might be 100 times larger than bacteria.

The classical geneticists and their academic successors never considered the possibility that a large chunk might be functionless, but rapid advancement in genomic studies in general and on the human genome in particular since 1964 refined the determining number of protein-encoding genes of human genomes to 0.5 million a drastic cut down from 6.5 million genes before initiation of human genome project (Graur, 2016). In 2001, prior to the publication of HGP, the draft report revealed that there were 26,000- 30,000 protein-encoding genes in the human genome (Lander et al, 2001). The completion of sequencing of euchromatic regions of the human genome in 2004 by International Human Genome Sequencing Consortium (2004) revealed that the number of protein-encoding genes in the human genome went down to 24,500. Consequently, the number of protein-encoding gene in human genome further dropped to 20,500 in 2007 (Clamp et al, 2007) and was 18, 877 in 2009 (Pruitt et al, 2009). This was substantiated later in 2014 by the genome biologists who studied the protein-encoding genes of human nuclear genome (Ezkurdia et al, 2014).

Contrary to earlier perception of the classical molecular genetics and genomics, contemporary genomicists observed that the number of genes in a genome of some plants and prokaryotic organisms were higher than in human species (Pertea and Salzberg, 2010). At the end of HGP, it was revealed that only 2% human genome contain protein-encoding genes, whereas, 98% of the human genome was found to possess no protein-encoding genes, whose function was not related to either phenotypic expression or regulating evolution to a certainty (Graur, 2016). Such observation helped a group of scientists to jump to an over-simplified conclusion that non-functional 98% of DNA should be recognized as Junk DNA (Hayden 2010). Whereas, the post-HGP follow-up studies as well the scientists of project ENCODE (2012) revealed that more than 80% genome was functional which was difficult to accept. In order to recognize most of the genome functions of any organism, there must be a number of conditions that needed to be fulfilled, such as indefinite size of population with negligible or little chance of deleterious mutations of genes, generations of the population which would be a very short cycle and these are merely theoretical considerations rather than realistic possibilities (Graur, 2016). In the regime of Anthropocene, where environmental pollution has induced higher rate of mutational changes in perennial plants like *Populous tremuloides* and animal species like *Homo sapiens*, who have fragmentary population size and longer life cycle (as well as reproductive cycle), the existence of genome with more than 80% functional state is just a speculation having no scientific background.

In order to figure out the functionality of genome, Graur (2016) tried to redefine the conceptual interpretation of "biological function" as "selected-effect function" with a historical context. It hypothesized that unique traits evolved out of the pre-existing traits and were driven by "selected-effect function" in course of successional cycles of reproduction. Basically, it explains the "causal role" (driving force or origin) of an evolutionary progress driven by subsequent evolution of trait (Millikan, 1984; Neander, 1991). Whereas, another group of evolutionary biologists theorized in favor of "causal-role" function, which is non-historical, having no relation to ancestral trait and is rather considered as an "ad hoc" device helping emergence of unique traits in the course of evolution (Cummins, 1975; Amundson and Lauder, 1994). Though it is difficult to either prioritize the historical "selected-effect function" in the gradual progress of evolution or promoting the concept of non-historical, instantaneous "Causal-role" in evolutionary changes (like mutation-induced changes), Doolittle et al (2014) considered the first option as "Function" and the latter one as "Activity." However, a critical review of function and action-induced evolutionary progress, by Graur et al (2013) ascertained that though there is a thin line of distinction between the two entities "selected-effect function" might have a "Causal-role action" but "causal-role action" was not found to be integrated with "selected-effect function." When functional role of genome is concerned as the driving evolution, two fundamental concerns the of genome help to move forward "evolution in action," elucidated by genome's "causal-role activity" and its cause (Brunet and Doolittle, 2014). Brunet and Doolittle (2014) further construed that in order to understand the evolutionary role of genomes, a clear understanding of function is needed of the genome involved (the particular constituent of genome that triggered the causal-role activity) and why does such constituent exists in the genome (which is elucidated by "selected-effect function").

According to contemporary perception, a number of evolutionary biologists tried to explain the genomic functionality in the backdrop of

"Natural selection" which gave rise to utopian ideas that, like other universal entities, the biological holy grail of life on the genome would remain "ever-functional" as the environment was absolutely free from oxidizing entities (no oxygen, no oxidation, no aging) and that passage of time, entropy and mutation (particularly the preponderance of deleterious mutations) would not affect the functional extent and level of genome (Graur et al, 2013). In the contemporary oxidized ambience of the biosphere, where ageing is the usual process in the life cycle of an organism, "genomic functionality" is regulated as well as protected by favor of natural selection, unless cyclic cascades of deleterious mutation alter the DNA sequences of the genome so often that there would not be any stability of phenotypic traits of species. There would be dreadful "evolutionary chaos" where it's difficult to define the identity of species or a group of organisms as a result of hypothetical emergence of a number of species in a short period of time and biological world would have to go through unique experience of inter-specific competition everyday. According to most evolutionary biologists, the genomic function is defined under "selected effect function," which is the integrated and decrypted functions of the DNA sequence in some locations of genome to be expressed in specific, encoded proteins that is under screening or is protected under natural selection. In order to understand the functionality, the distinction between "function" and "action" needs to be understood as being active does not necessarily mean physical exertion which could be considered as function. At the same time even while being functional, an organism might not be active throughout the day or long periods of time. Similarly, a biologically "active" genome might not necessarily be recognized as "functional" in terms of evolutionary progress.

However, according to biological activities, genome can be divided into three segments (Graur, 2016):

a. The genomic segments that are transcribed and translated.

b. The part of the genome segments that are transcribed but not translated.

c. The genomic segments that are not transcribed.

All the three segments are biologically "active" but constitute "functional" and "non-functional" elements in terms of evolutionary function. So, the "causal-role activity" needs to be studied separately and be considered to understand genomic activity rather than explaining the genomic function which could be defined by analyzing the "selected–effect function." A number of scientists did not find any reason to consider "functional role activity" as genomic function. For example, number of contemporary genomicists are of the opinion that an instance of "functional–role activity" of genome such as "low-level non-coding RNA transcription" would define the genomic functionality due to its inadequate role to put "evolution in action" (Kellis et al, 2014). In order to clear the biased perception of two extreme ideologies of functionality of genome a critical review is necessary:

a.   Genome with large stretches of DNA molecules having "majority of DNA with no protein en-coding potential" and its two overlapping sub-categories (Krams and Bromberg, 2013; Mehta et al, 2013):

a.   Non-coding DNA: The segment of genome that does not encode proteins.

b.   Junk-DNA: The segments of DNA that are neither functional nor deleterious.

b.   Genome with large stretches of DNA molecules having "majority of protein encoding potential" (ENCODE project consortium, 2012; Kellis et al, 2014).

Based on the critical review of the erstwhile perception of the functionality of genome in the distinct stretches of DNA found in different location of genomes, Graur et al (2015) broadly categorized its constituents into—"Functional DNA" and "Rubbish DNA," mainly on the basis of "selected-effect function" and the structural and functional entity of these two distinct categories of DNA needs to be elucidated here. The internal relationship of different categories of functional and non-functional DNA is presented in fig. 9.1.

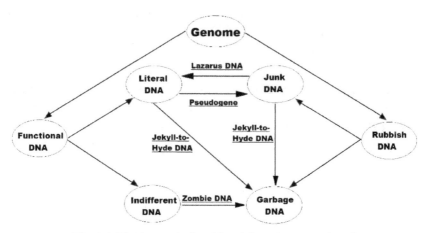

Fig. 9.1. The interrelationship of the sub-categories of
DNA in human genome (modified from Graur et al, 2015).

A. Functional DNA: It refers to the segment of the genome which is
    also recognized as "functional sequence." "Selected-effect function"
    of that is usually selected or maintained by "positive selection" or
    "balanced selection." The functional DNA is further categorized
    under two sub-categories:

a. Literal DNA: Under this category, the sequential order of the
    nucleotides which constitute the length of the literal DNA is found
    to be under selection. As an example of literal DNA, three
    combinations of triplet codons have been referred viz. ATT, ATA,
    and ATC that are responsible for encoding the amino acid
    Isoleucine among the 64 possible combinations of codon library
    (Graur, 2016). The literal DNA molecule comprises functional
    protein encoding genes, RNA-specifying genes and transcribed
    control elements of genome (Graur, 2016).

b. Indifferent DNA: In this category, the segments are functional but
    the sequential order of the nucleotides are not directly related to it.
    It prevents the DNA from being subjected to frame-shift mutations
    (served as fillers and spacers) and determines the size of nucleus
    and carry out nucelotypic functions. The scientists referred to the
    third-codon position of the degenerate codon systems as an ideal
    example of indifferent DNA because they have not been considered
    for a specific genomic functions. However, the absence of

nucleotide filler in this position would render the codon incomplete for functioning (Graur, 2016). Genomicists further theoresized that "indifferent DNA" selection has no regulating potential against or in favor of point mutation but plays a cardinal role in mutational changes like insertion or deletion in the molecule.

So, if we review the functional role of a literal DNA and an indifferent DNA as composite functions of a functional DNA, then both entities play an active role in driving evolutionary progress.

B. Rubbish DNA: Though this unusual terminology was used by Brenner (1998), the structural and functional existence is supported by contemporary geneticists of the 21st century. The "Rubbish DNA" was found to have no "selected-effect function" which perplexed scientists (Graur et al, 2015). In order to elucidate further, the functional and structural features of Rubbish DNA is subdivided into: "junk DNA" and "garbage DNA."

a. Junk DNA: This terminology was used by earlier cytologists and geneticists like Aronson et al (1960); Ehret and de Haller (1963), colloquially, in genetical studies in a broad sense (mostly relevant to studies of extra-chromosomal hereditary material). It was also used by Ohno (1972); Makalowski (2003) who said that Junk DNA's little or no involvement in evolutionary function is not impacted by selection (either in favor or against). Graur (2016), in his research communication mentioned that there is a conceptual resemblance of the "Junk DNA" to the junk in a storage as it could either contain the excess, less-used or un-used materials not required for daily usage or it could contain the article that is not used for current usage but could be useful in the future.

So, strategically, junk would not be recognized as a typically useless article (as opposed to garbage) as all the junk of the past or present may be useful items in the future (Graur, 2016). With reference to elucidating the identity of Junk DNA, earlier research of the molecular geneticists need to be mentioned as they don't believe that it played a role in evolution (Wen et al, 2012). So, without question it was held that if any piece of DNA (like Junk DNA) failed to prove its functionality, it would be considered as "non-functional." However, the nature of "evolutionary function" is quite complex to accept or discard in close-ended brackets. Rather, evolutionary functionality followed the Aristotelian doctrine of "potentiality' and "actuality" as a segment and has always been exposed to mutational changes in its oxidized environment where random alterations, whether "helpful" or "deleterious," would change its identity and functionality. The recognition of functional potential of Junk DNA has helped to recognize the occurrence of biological evolution in the living entity as empirical testimony of shift in paradigm that has driven the concerned living entity to move forward from its ancestral state to the derived state to be adapted in the dynamic environment, intermittently changed in passage of time, so it (i.e. junk DNA) might be regarded as the functional index of evolution for any biological organism, to ensure its existence in the race of "survival of the fittest." There is a saying "in order to see tomorrow, we need to live today." Similarly, if "Junk DNA" (with its huge treasure trove) would not exist, the evolutionary function which has ensured the emergence of new traits in the newly emerged organisms or derived organisms with unique traits would not be possible on the Earth.

b.   Garbage DNA: The term garbage DNA was distinguished from Junk DNA by Brenner (1998). Scientists hypothesized that the stretches of DNA sequence which were not selected, appeared detrimental to the sequence in the genome in the course of evolution. Therefore, this garbage DNA has little integration in the evolutionary process. However, the four sub-categories of DNA in the genome i.e. literal DNA, indifferent DNA, junk DNA and garbage DNA could broadly perform either of the evolutionary functions in the following three categories (Graur et al, 2015):

1. Transcribed and translated (TT)

2. Transcribed but not translated (Tt)

3. Not transcribed (nTt)

So, four categories of DNA in the genome could perform evolutionary function within 3 categories whose permutation and combination would lead up to 12 unique structural and functional entities such as literal DNA-Tt, junk DNA-nTt, garbage DNA-tt etc. Eminent molecular geneticist, Ohno

(1972) further elucidated these twelve structural and functional DNA entities of genome where dynamic environmental changes were found to play an important role in maintaining the unique identity of these DNA. Interestingly, genomicists witnessed a wide array of factors that changed the dynamic functional identity of these DNA from one form to another. It was noticed that the presence of pseudogenes helped to transform literal DNA to Junk DNA or occurrence of some disease-lead transformation of functional DNA to garbage DNA (Chen el al, 2003). The changes of evolutionary functional affiliation of the DNA led to further recognition of "Lazarus DNA" (when rubbish DNA transforms to functional DNA under mutational change), "Jekyll-to-Hyde DNA" (when functional DNA changes to garbage DNA under mutation) and "Zombie DNA" (when junk DNA mutated to garbage DNA) (Graur et al, 2015).

There is a grey area which questions whether such transformation of functional category of the DNA (from functional to non-functional or vice versa) happens at the time of evolutionary changes or leads to the evolutionary changes. However, the determination of dynamic functionality or non-functionality of the genomic segments have mainly been adjudged on the basis of "actual" as well as present status of evolutionary function rather than its "potential" to change in the future (Graur et al, 2015). Progress in advance studies of comparative genomics revealed that identification of the functional sites of the genome determined by comparing the degree of conservation between sequences from two or more species, helped to estimate the part of the functional nucleotides in the genome of human being which was in between 3-15% (Ward and Kellis, 2012). Scientists observed that gain or loss of functional elements during evolutionary changes needs to be understood by studying the ratio of functional nucleotides under selection against the backdrop of divergence of two or more species (Graur, 2016). As an example, the estimation of constraints between phylogenetically-related species would solely include the sequences (those contained by their predecessors) that should not be altered, lost or become non-functional and transferred to extant lineages of that ancestral predecessor (Graur, 2016). The gain or loss of the purifying selection in a particular locus of the genome is ascribed as "functional element turnover" as the composite action of changes, like physical ambiance and genetic profile, would lead to functional or non-functional state or vice versa (Graur, 2016).

Contemporary genomic investigations of humans revealed that around 8.2% of the genome is functional and inert areas like promoter region, transcription binding sites, untranslated regions, lncRNA s (long non-coding RNA) hardly help the functional fraction areas (Rands et al, 2014; Graur, 2016). It was further ascertained that lncRNA is devoid of any protein translation activity, rather, it played a critical role in expediting the conversion of functional element turnover of non-coding genomic areas, hence it was recognized as creator of "transcriptional noise" (Losick and Desplan, 2008; Urban and Johnston, 2018). The advancement in studies of the non-functioal regions of the human genome helped evolutionary biologists and genomicists determine that the slow pace of evolutionary action (corrective actions) through natural selection on diploid human genomes appeared to be functionally inefficient so as to get rid of the accumulation of deleterious alleles in the gene pool of the concerned population. So, there is substantial presence of garbage DNA in human genome, though, volume-wise, its existence is lesser in comparison to Junk DNA (Graur, 2016). There are a number of possibilities anticipated that are responsible for non-detection and inefficient eradication of these deleterious alleles, such as:

a. As the lesser frequency of these deleterious alleles are found to be present in the heterozygous state in the genome, it is not traced by the screening of natural selection.

b. Deleterious alleles are hardly detected/caught when passed down from generation to generation so a comparative examination between fixed alleles and polymorphic alleles (which could help to trace deleterious alleles) has not been performed.

A number of qualitative and quantitative studies have been conducted to determine the garbage genome in humans. In a recent investigation on 6515 exomes (individuals), scientists traced the origin of most protein-encoding genes, done by Fu et al (2013), and noticed that around 73% of the single-nucleotide variants and 86% of variants were found to be deleterious. Computer simulations helped to predict that all these deleterious variants evolved between 5000 and 10,000 years ago. It should also be mentioned that a unique genomic function was observed in the specific location of genome called SNP (Single Nucleotide polymorphism) where substitution of a single nucleotide led to a certain degree of variation in the human population. The occurrence of SNP was found in coding sequence, non-coding sequences and the space between two or more genes. SNP also leads to the occurrence of a number of carcinogenic diseases like thalassemia, cystic fibrosis etc. in humans (Chang and Kan, 1979; Hamosh et al, 1992). SNPs in the coding region of the genes was further categorized into two: Synonymous SNP, which does not alter original protein sequence, and Nonsynonymous SNP, which alters the amino acid sequence of the protein. The nonsynonymous SNP was found to be integrated in two distinct forms of point mutation called missense and nonsense mutations. Studies revealed that the human genome contained around 1000 SNVs (nonsynonymous single nucleotide variants) which is responsible for downgrading of fitness of human species in course of its journey of evolution (Sunyaev et al, 2001).

Based on the advancement in contemporary researches on molecular genetics, molecular phylogenetic and genomics, genome biologists are of opinion that genome duplication along with factors like sub-genomic duplication, replicative transpositions (of transposons), insertions of mononucleotides and oligonucleotides could play a major role in increasing genome size (gradual mutualism and punctuated equilibrium). Whereas, the decrease in genome size is by means of gradual mutualism (Piegu et al, 2006; Graur, 2016). The scientists further observed that such increase in size would not ensure the fractional increase of non-functional Junk DNA unless the newly-added functional DNA stretch becomes nonfunctional due to evolutionary functional glitch. The careful observation of scientists ascertained that transcription and reverse transcription is not an efficient process in copying genetic information and the mutational action induced by replicative transposition of transposons or jumping genes play a cardinal role in increasing genome size, and consequently, increase in non-functional elements or Junk DNA in the genome (Dietrich et al, 2004; Kovach et al, 2010).

It was observed that around 60% variation of the genome size is triggered by the content of transposons and was generally accepted that variation of non-functional sequences in the genome, specifically, in Junk DNA content of consecutive species (Tang, 2015). On the basis of critical review, scientists affirmed that large genome contains higher number of transposons and smaller genome contain lesser number of transposons and the existence of Junk DNA plays a cardinal role in coordination of evolutionary function in the biological world, where the concept of fitness has changed with the passage of time and space (Kidwell, 2002; Hawkins et al, 2006; Kovach et al, 2010; Rands et al, 2014). A number of hypothesis have been proposed from time to time to define the structural formation of large genome with relation to accumulation of DNA, specifically non-functional DNA, and "coincidence hypothesis" is one such hypothesis that could explain the structural formation of larger genome in organisms, mainly in animals. In "coincidence hypothesis," Pagel and Johnstone (1992) promulgated that bigger size of the cell is compatible for accumulation of large quantity of DNA that contains larger size of genome. According to this hypothesis, most of the accumulated DNA in the large-sized genome is recognized as Junk DNA where increased rate of mutational pressure regulate the increase in the size of genome. It was also explained that increasing of the genome size is not beyond the limit as its maintenance seem to create energy-limitation issues and was found to be a genetically deleterious process assumed to be selected against.   On   the   contrary "nucleotypic hypothesis" (Davis and Heywood, 1963) ascribed the entire genome as functional but it constituted of little "literal DNA" and large quantities of "indifferent DNA." This hypothesis promoted the idea that positive selection of large cells happened parallel to the changing dynamics of environment which corresponded to more accumulation of nuclear volume. It was assumed to be fulfilled by accumulation of indifferent DNA or by means of DNA condensation in the genome. In early 20th century, the genome biologists tried to explain the non-functionality of a big chunk of genome due to presence of redundant DNA which is closely related to functional genes (Darlington, 1937). The theory of redundancy was supported by the renowned genomicist Ohno (1972) who further contended that the lion's share of eukaryotic genome is occupied by Junk DNA. Furthermore, he construed that "gene duplication" played an important role to pass the stiff screening measure of natural selection. It was stated that in post-duplication phase, when "one copy" of the gene is engaged in original function (reproduction and maintaining the consistency of phenotypic profile from one generation to next), the other copy would accumulate mutational changes and transform to pseudogene (contained in the non-functional part of genome) and emerge as the 'beneficial gene' which would enhance fitness of the extant organisms or promote the evolution of unique varieties. So, along with issues like redundancy during gene and genome duplications leading to the formation of Junk DNA, a fair number of transposable elements, those that are degraded by mutations, add-up the sink of Junk DNA in the non-functional zone of genome. Susumu Ohno (1972) stated further:

> *The creation of every new gene must have been accompanied by many other redundant copies joining the rank of silent DNA base sequences....Triumphs as well as failures of nature's past experiments appear to be contained in our genome.*

However, contemporary observations by molecular geneticists and genomicists show that huge gatherings of the transposons (those degraded in the course of mutations) in non-functional zone of genome along with its other non-functional allies such as pseudogenes, introns, repetitive DNAs support the Junk DNA hypothesis (Graur, 2016). In support of identical thematic proposition with non-functional entity of genome, another group of scientist came up with a unique hypothesis known as "Selfish DNA hypothesis" (Doolitile and Sapienza, 1980), that defined:

1. The DNA sequence produce and retain extra copy of itself in the genome.

2. The copied DNA sequence does not play any role in enhancing the fitness of the host organism; and is often found to be detrimental for host organism.

The recent investigations ascertained that selfish DNA is constituted of class-I and class-II transposons. The most unique mode of evolutionary action of the Selfish DNA opens up a new path of journey down for evolutionary science where a recent experimental observation by Daniels et al (1990) must be referred to. The scientists revealed that a transposable element known as "*p* element of *Drosophila*" is transferred from one genome to another to be colonized by HGT or "Horizontal gene transfer"[12].

---

[12] Horizontal gene transfer (HGT): This is a unique mode of transfer of the genetic material, other than the usual process of transfer of genetic material from parent to offspring (which has been recognized as vertical gene transfer) (Keeling and Palmer, 2008). Bacteriophage, plasmids, transposons etc. act as functional vector for expediting HGT (Varga et al, 2016). HGT plays an important role in the evolutionary progress of a number of organisms. It helps to evolve antibiotic-resistant bacteria and insecticide-resistant pests.

The genomicists noticed that HGT-driven logistics of transposable elements emigrate from one genome to colonize a new genome and its ecesis either helps it to escape unavoidable extinction due to the destruction of its hosts or its stochastic inactivation due to mutational degradation (Hart et al, 2004; 2009). On the other hand accumulation of "Selfish DNA" in non-functional part of the genome would help host species to cope with stochastic level of fitness in the backdrop of its dynamic environment. So, accumulation of Selfish DNA in the genome might be considered as "enrichment of the genomic junkyard" but the primary goal of the evolutionary functions is to meet the standards of fitness in terms of natural selection (either upgraded or downgraded by any unpredictable change of environment) where this genomic junkyard could be the last resort for a species to evade inevitable extinction.

The advancement of theoretical perceptions along with immense progress in molecular genetics, genetic engineering and genomics in the last two decades helped to give qualitative ideas on the evolutionary role of the non-functional segment of the genome. The existence of non-coding stretches of genome, Junk DNA and Dark DNA has facilitated exploration in evolutionary biology. Still the interpretation of the evolutionary functions of the darker part of the genome remained hidden as overall findings of genomic studies failed to reveal the functionaries of the common gene which plays a cardinal role in occurrence of human maladies. In support of the above statement, scientists confirmed that around *"three dozen of specific genetic variants have been associated with type 2 diabetes...but together, they have been found to explain 10 percent of the disease's heritability—the proportion of variation in any given trait that can be explained by genetics"* (Hall, 2010). The genome biologists and epigeneticists opined that a number of human maladies such as cardiac ailments, high blood pressure, schizophrenia have identical "missing heritability" issues (Hall, 2010). With reference to genetic issues like "missing heritability," Joseph Nadeau, the head of Genetics at Western Reserve University of Cleveland, USA, along with his co-worker Vicki Nelson (Nelson et al, 2010) experimentally created evolved, inbred mice with the traits that were regulated by the variant DNA found in the genome of their parental lineages but, surprisingly, missing in the genomes of the inbred mice themselves. It is a matter of surprise that the un-inherited DNA of the previous generations have very intense impact such that expression of phenotypic traits of the missing text would last for a few generations. Such trans-generational genetic effect of inheritance was also found equally important to the conventional Mendelian principle of inheritance in biological organisms (Nelson et al, 2010). So, Nadeau's experiment revealed that expression of a particular gene depended upon the presence of surrounding genes and such interaction could manifest in the form of unique traits which might appear as phenotypic expressions in the offsprings without being present in parental lineages. In the language of Nadeau (Hall, 2010), "situational genetics" is where variant DNA in a genome could influence the molecular neighborhood of the genes and the expression of the unique trait might be witnessed in the offspring for a number of consequent generations in the absence of its inherent genotypic entity. Such "trans-generational genetic effect" could help trace the "missing–heritability problem" buried under the conventional Mendelian principle of inheritance hidden in the genome's dark matter. In some sense, "trans-generational genetic effect" has evolved as the contemporary, a thematic treasure trove in the field of epigamic studies in the 21st century.

To move one step further, Man-Yee Lam, one of Nadeau's research student, tried to demonstrate the trans-generational effect in mice so as to identify the interactions of "modifier" genes which might enhance the susceptibility of testicular germ-cell tumor (Hall, 2010). At the end of the research, Lam successfully analyzed the results of the control and the newly-bred mice and found positive interactions with "modifier" genes in relation to gene mutation susceptibility of this particular type of testicular carcinogens. Lam and her colleagues also noticed that the control lineages of the inbred mice population did not have any earlier symptoms of disease mutations that showed higher rate of testicular cancer. Thus, they came to the conclusion that possession of genetic variant, creating a risk of testicular cancer, pre-existed in the genome of parental and grandparental lineages of the newly-bred populations. It showed that the "multitude of genes" as well as "genetic neighborhoods" played a cardinal role in expression of certain traits in the offsprings for a number of generations without passage of genetic variants from parent to offspring (Hall, 2010). Hence, trans-generational effect of genes and family history of the parental lineage need to be reviewed to retrieve variant genes responsible for a number of traits and genetically-inherited disease normally contained by the respective genome of the concerned organism. Such dark patches of DNA (variant genes) might not be inherited from parental generation and affect the traits of the children, according to the principles of "trans-generational genetic effect."

In order to find out whether genomes of human beings would experience "gene to gene interplay" as well as be accustomed to "trans-generational genetic effect," Eric J. Topol, the head of Genomic Research in Scripps Research Institute, California, USA, carried out studies on 80-year-old subjects (having no record of medications for ailments) and explained that all the disease-free people had not inherited the variant genes or "risk alleles," causing genetically-inherited disease to be countered by pathological mutation (Hall, 2010). As per Mendelian principle of inheritance, the subjects, possessing "risk alleles,' are supposed to be more susceptible to the disease but realistically possession of "risk alleles" would not let the disease develop (Hall, 2010). Eric Toprol came up with an explanation which stated that the development of genetically-inherited disease does not solely depend upon the gene action of disease variant gene but on the cumulative effect of other gene variants in its vicinity which might counteract or mask the harmful effect of that disease-causing variant gene (Hall, 2010). So, according to Toprol (Hall, 2010), it's not the sole action of constituents like "risk alleles" or "disease-causing variant gene" in the gene and genome that determine the expression of certain traits, but the cumulative and counteractive effect of "protective alleles," "modifier genes" with "disease-causing variant gene" as well as "risk alleles" which redefine the development of disease and its evolutionary progress. Most importantly if we don't recognize the genomic junkyard as the genomic treasure trove of evolutionary instruments, like modifier genes, protective alleles etc. the narrative of evolutionary function of biological organisms, Selfish DNA, Junk DNA or Dark DNA would remain non-scientific jargon for us and our future generations.

When Eshna Gogia used the term "Junkomics" in his blog titled "Junkomics: Digging' through genomic 'garbage' for the hidden treasure," published on 26th December, 2017, it seemed for "geeks" or "nerds." However, a brief review of the article shed light on issues like potential threats of applying medicine against bacterial strains, resistance against bacteria or application of a number of pesticides which were found to be resistant against a number of pests. A number of genomicists have tried to explore the uncharted territory of "Junk genome" which has resonated in the firm affirmation of the famous genomicist, Pawan K. Dhar, who said,

"Given alarming resistance emerging towards existing drugs, we need out-of-the-box solutions for effective treatment of disease." (Gogia, 26th December, 2017) So, future studies on Junkomics, particularly the dark segment of genome, is the ultimate dream that future genomicists need to explore. The proverb "exception that proves the rule" is still valid for anything and everything regarding substantiating any scientific principle and establishing any scientific principle of genomics. Nadeau and his fellow genomicists traversed the uncharted territory of evolution to discover an alternative principle of inheritance and strived to elucidate the unique model of "genome-wide association studies" which supplemented Mendelian principles of inheritance. In support of proving the efficacy of "trans-generational effect of gene" as a supplementary principle of Mendelian principle of inheritance, Nadeau (Hall, 2010) had said, *"Mendel picked the traits where he would get simple genetics....What Mendel said is true...but it's not the whole truth."*

References

Amundson, R., and Lauder, G.V. (1994). Function without purpose. Biol Philos 9: 443–469.
https://doi.org/10.1007/BF00850375

Aronson, A.I., Bolton, E.T., Britten, R.J., Cowie, D.B., Duerksen, J.D., et al... (1960). Biophysics. Year book - Carnegie Institution of Washington (1960). Volume 59. Baltimore, MD: Lord Baltimore Press. pp. 229–289.

Brenner S. (1998). Refuge of spandrels. Curr Biol. 8(19):R669. doi: 10.1016/s0960-9822(98)70427-0.

Brunet, T.D., and Doolittle, W.F. (2014) Getting "function" right. PNAS U S A. 111(33): E3365. doi: 10.1073/pnas.1409762111.

Chang, J.C. and Kan, Y.W. (1979). β^0 Thalassemia, a Nonsense Mutation in Man. PNAS USA, 76: 2886-2889.
http://dx.doi.org/10.1073/pnas.76.6.2886

Chen, Y. H., Xu, S.J., Bendahhou, S., Wang, X.L., Wang, Y., Xu, W.Y., Jin, H.W., Sun, H., Su, X.Y., Zhuang, Q.N., Yang, Y.Q., Li, Y.B., Liu, Y., Xu, H.J., Li, X.F., Ma, N., Mou, C.P., Chen, Z., Barhanin, J., and Huang, W. (2003). KCNQ1 gain-of-function mutation in familial atrial fibrillation. Science. 299(5604):251-4. doi: 10.1126/science.1077771.

Clamp, M., Fry, B., Kamal, M., Xie, X., Cuff, J., Lin, M.F., Kellis, M., Lindblad- Toh, K., and Lander, E.S. (2007). Distinguishing protein-coding and noncoding genes in the human genome. PNAS U S A. 104(49):19428-33. doi: 10.1073/pnas.0709013104. Epub 2007 Nov 26.

Cummins, R. (1975). 'Functional Analysis.' Journal of Philosophy, LXXII:741-65.

Daniels, D., Zuber, P., and Losick, R. (1990). Two amino acids in an RNA polymerase sigma factor involved in the recognition of adjacent base pairs in the -10 region of a cognate promoter. PNAS U S A. 87(20): 8075-9. doi: 10.1073/pnas.87.20.8075.

Darlington, C. D. (1937). Recent Advances in Cytology, 2nd ed. J. and A. Churchill, London.

Davis, P.H. and Heywood, V.H. (1963). Principle of Angiosperm Taxonomy. Oliver & Boyd, London, P. 556.

Dietrich, F.S., Voegeli, S., Brachat, S., Lerch, A., Gates, K., Steiner, S., Mohr, C., Pöhlmann, R., Luedi, P., Choi, S., Wing, R.A., Flavier, A., Gaffney, T.D., and Philippsen, P. (2004). The *Ashbya gossypii* genome as a tool for mapping the ancient Saccharomyces cerevisiae genome. Science. 304(5668): 304-7. doi: 10.1126/science.1095781. Epub 2004 Mar 4.

Doolittle, W.F., Brunet, T.D.P., Linquist, S. and Gregory, T.R. (2014). Distinguishing between "Function" and "Effect" in Genome Biology. *Genome Biology and Evolution.* 6(5): 1234–1237.
https://doi.org/ 10.1093/gbe/evu098

Doolittle, W., and Sapienza, C. (1980). Selfish genes, the phenotype paradigm and genome evolution. Nature 284: 601–603. https://doi.org/10.1038 /284601a0

Ehert,C.F., and de Haller, G. (1963). Origin, development, and maturation of organelles and organelle systems of the cell surface in Paramecium. J Ultrastruct Res 23: SUPPL 6: 1–42.

Ezkurdia, I., Juan, D., Rodriguez, J.M., Frankish, A., Diekhans, M., Harrow, J., Vazquez, J., Valencia, A., and Tress, M.L. (2014). Multiple evidence strands suggest that there may be as few as 19,000 human protein-coding genes. Hum Mol Genet. 23(22):5866-78. doi: 10.1093/hmg/ddu309.

Fu, Y., Foden, J.A., Khayter, C., Maeder, M. L., Reyon, D., Joung, J. K., and Sander, J. D. (2013). High-frequency off-target mutagenesis induced by CRISPR-Cas nucleases in human cells. Nat Biotechnol. 31(9):822-6. doi: 10.1038/nbt.2623.

Gogia, E. (26th December, 2017). Junkomics. Diggin' through genomic 'garbage'for the hidden treasure. https://nucleicacidmemory.com/junkomics -diggin-through-genomic-garbage-for-the-hidden-treasure-d34bfc0e3926

Graur, D. (2016). Molecular and genome evolution. Sunderland (MA): Sinauer Associates.p.612.

Graur, D., Zheng, Y., and Azevedo, R. B. (2015). An evolutionary classification of genomic function. Genome Biol Evol. 7(3): 642-5. doi: 10.1093 /gbe /evv021.

Graur, D., Zheng, Y., Price, N., Azevedo, R.B., Zufall, R.A., and Elhaik, E. (2013). On the immortality of television sets: "function" in the human genome according to the evolution-free gospel of ENCODE. Genome Biol Evol. 5(3): 578-90. doi: 10.1093/gbe/evt028.

Hall, B. K. (2010). The Neural Crest and Neural Crest Cells in Vertebrate Development and Evolution. Springer. P. 402.

Hamosh, A., King, T.M., Rosenstein, B.J., Corey, M., Levison, H., Durie, P., Tsui, L.C., McIntosh, I., Keston, M., Brock, D. J., Macek, M., Zemková, D., Krásničanová, H., Vávrová, V, Golder, N., Schwarz, M.J., Super, M., Watson, E.K., Williams, C., Bush, A., O'Mahoney,S.M., Humphries, P., DeArce, M.A., Reis, A., Bürger, J., Stuhrmann, M., Schmidtke, J., Wulbrand, U., Dörk, T., Tümmler, B., and Cutting G.R. (1992). Cystic fibrosis patients bearing both the common missense mutation Gly---- Asp at codon 551 and the delta F508 mutation are clinically indistinguishable from delta F508 homozygotes, except for decreased risk of meconium ileus. Am J Hum Genet. 51(2):245-50.

Hart, P.S., Becerik, S., Cogulu, D., Emingil, G., Ozdemir-Ozenen, D., Han, S.T., Sulima, P.P., Firatli, E., and Hart, T.C. (2009). Novel FAM83H mutations inTurkish families with autosomal dominant hypocalcified amelogenesis imperfecta. Clin Genet. 75(4): 401-4. doi: 10.1111/j.1399- 0004.2008. 01112.x. Epub 2009.

Hart, P.S., Hart, T.C., Michalec, M.D., Ryu, O.H., Simmons, D., Hong, S., Wright, J.T. (2004). Mutation in kallikrein 4 causes autosomal recessive hypomaturation amelogenesis imperfecta. J Med Genet. 41(7):545-9. doi: 10.1136/jmg.2003.017657.

Hawkins, J. S., Kim, H., Nason, J. D., Wing, R. A., and Wendel, J. F. (2006). Differential lineage-specific amplification of transposable elements is responsible for genome size variation in Gossypium. Genome Res. 16 (10): 1252-61. doi: 10.1101/gr.5282906. Epub 2006 Sep 5.

Hayden, S., Bekaert, M., Crider, T.A., Mariani, S., Murphy, W.J., and Teeling, E.C. (2010). Ecological adaptation determines functional mammalian olfactory subgenomes. Genome Res 20(1):1-9. doi: 10.1101/gr.099416.109. Epub 2009 Dec 1. PMID: 19952139

International Human Genome Sequencing Consortium. (2004). Finishing the euchromatic sequence of the human genome. Nature 431: 931–945. https://doi.org/10.1038/nature03001

Keeling, P. J., and Palmer, J. D. (2008). Horizontal gene transfer in eukaryotic evolution. Nat Rev Genet. 9(8): 605-18. doi: 10.1038/nrg2386.

Kellis, M., Wold, B., Snyder, M.P., Bernstein, B.E., Kundaje, A., Marinov, G.K., Ward, L.D., Birney, E., Crawford, G.E., Dekker, J., Dunham, I., Elnitski, L.L., Farnham, P.J., Feingold, E.A., Gerstein, M., Giddings, M.C., Gilbert, D.M., Gingeras, T.R., Green, E.D., Guigo, R., Hubbard, T., Kent, J., Lieb, J.D., Myers, R.M., Pazin, M.J., Ren, B., Stamatoyannopoulos, J.A., Weng, Z., White, K.P., and Hardison, R.C. (2014). Defining functional DNA elements in the human genome. PNAS U S A. 111(17): 6131-8. doi: 10.1073/pnas.1318948111. Epub 2014 Apr 21.

Kidwell, M.G. (2002). Transposable elements and the evolution of genome size in eukaryotes. Genetica 115: 49–63. https://doi.org/10.1023/A: 1016072014259

Kovach, A., Wegrzyn, J.L., Parra, G., Holt, C., Bruening, G.F., Loopstra, C.A., Hartigan, J., Yandell, M., Langley, C.H., Korf, I., and Neale, D.B. (2010). The *Pinus taeda* genome is characterized by diverse and highly diverged repetitive sequences. BMC Genomics 11: 420 (2010). https://doi.org/10. 1186/1471-2164-11-420

Krams, S.M., and Bromberg, J.S. (2013). Encode: life, the universe, and everything. Ann Neurol. 73(5): A8-9. doi: 10.1002/ana.23920.

Lander, E.S., Linton, L.M., Birren, B., Nusbaum, C., Zody, M.C., Baldwin, J., Devon, K., Dewar, K., Doyle, M., FitzHugh, W., Funke, R., Gage, D., Harris, K., Heaford, A., Howland, J., Kann, L., Lehoczky, J., LeVine, R., McEwan, P., McKernan, K., Meldrim, J., Mesirov, J.P., Miranda, C., Morris, W., Naylor, J., Raymond, C., Rosetti, M., Santos, R., Sheridan, A., Sougnez, C., Stange-Thomann, Y., Stojanovic, N., Subramanian, A., Wyman, D., Rogers, J., Sulston, J., Ainscough, R., Beck, S., Bentley, D., Burton, J., Clee, C., Carter, N., Coulson, A., Deadman, R., Deloukas, P., Dunham, A., Dunham, I., Durbin, R., French, L., Grafham, D., Gregory, S.,

Hubbard, T., Humphray, S., Hunt, A., Jones, M., Lloyd, C., McMurray, A., Matthews, L., Mercer, S., Milne, S., Mullikin, J.C., Mungall, A., Plumb, R., Ross, M., Shownkeen, R., Sims, S., Waterston, R.H., Wilson, R.K., Hillier, L.W., McPherson, J.D., Marra, M.A., Mardis, E.R., Fulton, L.A., Chinwalla, A.T., Pepin, K.H., Gish, W.R., Chissoe, S.L., Wendl, M.C., Delehaunty, K.D., Miner, T.L., Delehaunty, A., Kramer, J.B., Cook, L.L., Fulton, R.S., Johnson, D.L., Minx, P.J., Clifton, S.W., Hawkins, T., Branscomb, E., Predki, P., Richardson, P., Wenning, S., Slezak, T., Doggett, N., Cheng, J.F., Olsen, A., Lucas, S., Elkin, C., Uberbacher, E., Frazier, M., Gibbs, R.A., Muzny, D.M., Scherer, S.E., Bouck, J.B., Sodergren, E.J., Worley, K.C., Rives, C.M., Gorrell, J.H., Metzker, M.L., Naylor, S.L., Kucherlapati, R. S., Nelson, D. L., Weinstock, G. M., Sakaki, Y., Fujiyama, A., Hattori, M., Yada, T., Toyoda, A., Itoh, T., Kawagoe, C., Watanabe, H., Totoki, Y., Taylor, T., Weissenbach, J., Heilig, R., Saurin, W., Artiguenave, F., Brottier, P., Bruls, T., Pelletier, E., Robert, C., Wincker, P., Smith, D.R., Doucette-Stamm, L., Rubenfield, M., Weinstock, K., Lee, H.M., Dubois, J., Rosenthal, A., Platzer, M., Nyakatura, G., Taudien, S., Rump, A., Yang, H., Yu, J., Wang, J., Huang, G., Gu, J., Hood, L., Rowen, L., Madan, A., Qin, S., Davis, R.W., Federspiel, N.A., Abola, A.P., Proctor, M.J., Myers, R.M., Schmutz, J., Dickson, M., Grimwood, J., Cox, D.R., Olson, M.V., Kaul, R., Raymond, C., Shimizu, N., Kawasaki, K., Minoshima, S., Evans, G.A., Athanasiou, M., Schultz, R., Roe, B.A., Chen, F., Pan, H., Ramser, J., Lehrach, H., Reinhardt, R., McCombie, W.R., de la Bastide, M., Dedhia, N., Blöcker, H., Hornischer, K., Nordsiek, G., Agarwala, R., Aravind, L., Bailey, J.A., Bateman, A., Batzoglou, S., Birney, E., Bork, P., Brown, D.G., Burge, C.B., Cerutti, L., Chen, H.C., Church, D., Clamp, M., Copley, R.R., Doerks, T., Eddy, S.R., Eichler, E.E., Furey, T.S., Galagan, J., Gilbert, J.G., Harmon, C., Hayashizaki, Y., Haussler, D., Hermjakob, H., Hokamp, K., Jang, W., Johnson, L.S., Jones, T.A., Kasif, S., Kaspryzk, A., Kennedy, S., Kent, W.J., Kitts, P., Koonin, E.V., Korf, I., Kulp, D., Lancet, D., Lowe, T.M., McLysaght, A., Mikkelsen, T., Moran, J.V., Mulder, N., Pollara, V.J., Ponting, C.P., Schuler, G., Schultz, J., Slater, G., Smit, A.F., Stupka, E., Szustakowski, J., Thierry-Mieg, D., Thierry-Mieg, J., Wagner, L., Wallis, J., Wheeler, R., Williams, A., Wolf, Y.I., Wolfe, K.H., Yang, S.P., Yeh, R.F., Collins, F., Guyer, M.S., Peterson, J., Felsenfeld, A., Wetterstrand, K.A., Patrinos, A., Morgan, M.J., de Jong, P., Catanese, J.J., Osoegawa, K., Shizuya, H., Choi, S., Chen, Y.J., Szustakowki, J., and International Human Genome Sequencing Consortium. (2001). Initial sequencing and analysis of the human genome. Nature. 409(6822):860-921. doi: 10.1038 /35057062. Erratum in: Nature 2001 Aug 2;412(6846):565. Erratum in: Nature 2001 Jun 7;411(6838):720. Szustakowki, J [corrected to Szustakowski, J]. PMID: 11237011.

Losick, R., and Desplan, C. (2008). Stochasticity and cell fate. Science. 320(5872): 65-8. doi: 10.1126/science.1147888.

Makalowski, W. (2003). Genomics. Not junk after all. Science. 300(5623): 1246- 7. doi: 10.1126/science.1085690.

Mehta, D., Klengel, T., Conneely, K. N., Smith, A. K., Altmann, A., Pace, T. W., Rex-Haffner, M., Loeschner, A., Gonik, M., Mercer, K. B., Bradley, B., Müller-Myhsok, B., Ressler, K. J., and Binder, E. B. Childhood maltreatment is associated with distinct genomic and epigenetic profiles in posttraumatic stress disorder. PNAS U S A. 110(20): 8302-7. doi: 10.1073/pnas.1217750110. Epub 2013 Apr 29.

Millikan, R. (1984). Language, thought, and other biological categories. Cambridge: MIT.

Neander, K. (1991). The teleological notion of function. Australasian Journal of Philosophy. 69:454–68.

Nelson, V.R., Spiezio, S.H., and Nadeau, J.H. (2010). Transgenerational genetic effects of the paternal Y chromosome on daughters' phenotypes. Epigenomics. 2(4): 513-21. doi: 10.2217/epi.10.26.

Ohno, S. (1972). So much "junk" DNA in our genome. In: Smith. H. H., editor. Evolution of Genetic Systems. New York: Gordon and Breach. pp. 366–370.

Pagel, M., and Johnstone, R. A. (1992). Variation across species in the size of the nuclear genome supports the junk-DNA explanation for the C-value paradox. Proc Biol Sci. 249(1325): 119-24. doi: 10.1098/rspb.1992.0093. PMID: 1360673.

Pertea, M., and Salzberg, S. L. (2010). Between a chicken and a grape: estimating the number of human genes. Genome Biol. 11(5):206. doi: 10.1186/gb- 2010-11-5-206. Epub 2010 May 5.

Piegu, B., Guyot, R., Picault, N., Roulin, A., Sanyal, A., Kim, H., Collura, K., Brar, D.S., Jackson, S., Wing, R.A., and Panaud, O. (2006). Doubling genome size without polyploidization: dynamics of retrotransposition-driven genomic expansions in *Oryza australiensis*, a wild relative of rice. Genome Res. 16(10):1262-9. doi: 10.1101/gr.5290206. Epub 2006 Sep 8. Erratum in: Genome Res. 2011 Jul;21(7):1201. Saniyal, Abhijit [corrected to Sanyal, Abhijit].

Pruitt, K.D., Tatusova, T., Klimke, W., Maglott, D.R. (2009). NCBI Reference Sequences: current status, policy and new initiatives. Nucleic Acids Res. 37(Database issue):D32-6. doi: 10.1093/nar/gkn721. Epub 2008 Oct 16.

Rands, C.M., Meader, S., Ponting, C.P., and Lunter, G. (2014). 8.2% of the Human genome is constrained: variation in rates of turnover across functional element classes in the human lineage. PLoS Genet. 10(7): e1004525. doi: 10.1371/journal.pgen.1004525.

Sunyaev, S., Ramensky, V., Koch, I., Lathe, W 3rd., Kondrashov, A.S., and Bork, P. Prediction of deleterious human alleles. (2001). Hum Mol Genet. 10(6): 591-7. doi: 10.1093/hmg/10.6.591.

Tang, D. (2015). Repetitive elements in vertebrate genomes. http://davetang.Org /muse/2014/05/22/repetitive-elements-in-vertebrate-genomes/

The ENCODE Project Consortium. (2012). An integrated encyclopedia of DNA elements in the human genome. Nature 489: 57–74. https://doi.org /10. 1038/nature11247

Urban, E.A. and Johnston, R. J. Jr. (2018). Buffering and Amplifying Transcriptional Noise During Cell Fate Specification. Front. Genet. 9: 591. doi: 10.3389/fgene.2018.00591

Varga, T., Mounier, R., Patsalos, A., Gogolák, P., Peloquin, M., Horvath, A., Pap, A., Daniel, B., Nagy, G., Pintye, E., Póliska, S., Cuvellier, S., Larbi, S.B., Sansbury, B.E., Spite, M., Brown, C.W., Chazaud, B., Nagy, L. (2016). Macrophage PPARγ, a Lipid Activated Transcription Factor Controls the Growth Factor GDF3 and Skeletal Muscle Regeneration. Immunity. 45(5): 1038-1051. doi: 10.1016/j.immuni.2016.10.016. Epub 2016 Nov 8.

Vogel F. (1964). A preliminary estimate of the number of human genes. Nature. 201:847. Doi: 10.1038/201847a0.

Ward, L.D., and Kellis, M. (2012). Interpreting noncoding genetic variation in complex traits and human disease. Nat Biotechnol. 30(11): 1095-106. doi: 10.1038/nbt.2422.

Wen, R., Li, F., Xie, Y., Li, S., and Xiang, J. (2012). A homolog of the cell apoptosis susceptibility gene involved in ovary development of Chinese shrimp *Fenneropenaeus chinensis*. Biol Reprod. 86(1): 1-7. doi: 10.1095/biolreprod.111.092635.

Chapter 10

Evolutionary Balance-sheet of Human Evolution: Gradual loss of DNA (of Human Genome) at what cost?

Geneticists have discovered that modern humans actually possesses far less genetic information in our cells than their ancestral cousins. They estimate since early humans split from common ancestor we shared with our closest living relatives, chimpanzees, we have lost 40.7 million base pairs. These basic biochemical units are what make up DNA strands and group together to encode genes. Researchers say around half of these genetic sequences appear to have been repeated sections of DNA, but 27.96 million base pairs lost were unique...They found humans appear to have lost around 15.8 million base pairs after separating from apes early in our evolutionary history in Africa, around 13 million years ago. As humans then dispersed and spread around the world, they shed a further 12.16 million unique pieces of DNA.

---- Richard Gray

The progress of science either in the form of unique discoveries or resolution of long-drawn conflicts has always gone through the process of stochasticity rather than a deterministic path of certainty. Deciphering the emergence of human species on Earth through encrypted DNA, nature's holy grail, and stored in the human genome, an archive of evolution of modern human species, has had a long history of debatable interpretation of scientific observations and hypotheses since the 18th century.

Around 1955-56, two famous cytogeneticists, Joe Hin Tijo and Albert Levan of University of Lund, Sweden, were successful in developing a culture protocol of human fibroblasts which helped them to demonstrate karyotype analysis (diploid chromosome numbers) in the metaphase stage of cell division of the human genome. They found 46 chromosomes (i.e. 23 pairs) rather than 48 which was predicted earlier (Harper, 2006; Harper et al,

2007). The identical cytological preparation of apes phylogenetically close to the human species (chimpanzees, orangutans, bonobos, gorillas, etc.) revealed that all apes possess 48 chromosomes in their genomes (Tijo and Levan, 1956). Though, there was dearth of advance cytogenetical techniques such as chromosomal banding analysis, etc. the previous observation perplexed scientists as to whether human species fissured out of the ancestral mother stock of apes and if human species are a descendant of great apes, what was the tentative geologic period in which modern human species emerged as a separate species from the great-apes. After 10 years, geneticists were able to study the chromosomal rearrangements and advancement of DNA analysis in the chromosomal structures revealed that fusion of two chromosomes in apes, the phylogenetic cousins of the human species, yielded chromosome number 2 in the human genome. As the genetic content of the chromosome number 2 was found similar to the fused chromosomes of apes and which supposedly took place as a result of loss and rearrangement of genetic material (DNA sequence) between the two fused chromosomes at telomere regions (Ijdo et al, 1991).

The advanced level of karyotype analysis, supplemented with contemporary analysis of DNA sequencing of the extinct, phylogenetic predecessor of modern human species like the Denisovans and the Neanderthals revealed that the fused state of chromosome number 2 was also found in extinct predecessors of the modern human species (Meyer et al, 2012). It helped scientists to determine the tentative time of phylogenetic split of human species from greater apes, which supposedly took place around 0.75 to 4.5 million years ago (Rejon, 17th January, 2017). Through a comparative study of the genomic analysis among 125 human populations (out of random sampling of DNA), phylogeneticists and genomicists across the world were able to map the genetic diversity of modern human species (Gray, 7th August, 2015). They noticed that at the time of phylogenetic split from hominoids (i.e. greater ape lineage), modern human species lost around 15.8 million base pairs; the experimental data confirmed that the divergence of modern human species occurred in Africa and happened precisely around 13 million years ago (Gray, 7th August, 2015).

Contemporary studies on the comparative genomic analysis by different molecular biologists, genomicists and evolutionary biologists revealed that the chromosomal orientation in Denisovans, Neanderthals and modern human species indicated that fertile and viable descendants of the modern human species is the most likely explanation of evolution of hominids having 46 chromosomes each. However, hypothetical experiment of Mendelian inter-crossing between hominids (with 46 chromosomes) and hominoids (with 48 chromosomes) would create a tentative incompatibility of chromosomal pairing which is not revealed in any human genome studies. It does not have any supporting evidence that modern human genome possess genetic characteristic of greater apes in it (Rejon, 17th January, 2017). It further elucidates that chromosomal fusion in hominoids helped hominids evolve with reproductive isolation which acted as an evolutionary safeguard to develop an "efficient mechanism for reproductive isolation" that protected the modern human species from mixing with the ancestral lineages of big apes. However, fusion of two chromosomal entities must be associated with theoretical possibilities of rearrangement of DNA sequences and loss of a number of genes. In terms of origin and evolution of primates, mainly, the human species on Earth, such loss of genes should be recognized as an evolutionary gain where the most intelligent species on earth is concerned. Alternatively, the comparative genomic analysis proved that the modern human species possessed lesser genetic material than its "ancestral cousins" (Gray, 7th August, 2015).

Contemporary, quantitative estimation by genome biologists ascertained that as a result of phylogenetic split from the common ancestral lineage, specifically, the divergence from the closest cousin, chimpanzees, and human species has lost around 40.7 million base pairs (Gray, 7th August, 2015). Furthermore, scientists revealed that about 50% of the total base pair contains repetitive stretches of DNA in the human genome but modern human supposedly has lost around 27.96 million unique base pairs, rendering few thousand genes missing (Gray, 7th August, 2015). Contemporary genomicists observed that in the era of evolution (divergence) and diversification (adaptation followed by settlement of varied geographical race in different parts of the world), modern human lost almost 12.16 million pieces of DNA stretches (Gray, 7th August, 2015). Supporting the hypothesis of loss of genetic material by ancient human species, renowned geneticist Evan Eichler of University of Washington, USA, found that the first effort of the ancient human beings to venture "out of Africa' in order to colonize other parts of the world led to a loss of human population thereby, shrinkage of the population size led to substantial loss of genetic material (Gray, 7th August, 2015).

In comparative genomic analysis of the ancient human species (like Neanderthals and Denisovans) and modern human species, scientists noticed that the ancient human species possessed 104,000 unique base pairs, which were absent in the genome of modern human species. It indicated that modern human species lost these unique base pairs during the 'out of Africa' migration (Gray, 7th August, 2015). On the other hand, genome of modern human species possessed around 33,000 unique base pairs that were found in Neanderthals and Denisovans, which proved that either the ancient human species lost these unique base pairs earlier or the divergence of modern human species went through evolutionary diversification of new genes, resulting in mutational changes, and emerged with a number of unique base pairs (Gray, 7th August, 2015). According to contemporary statistical records of human genome studies, there are around 3 billion base pairs in 23 chromosomes of each somatic cell. Scientists further assessed that the average gene in the human genome is made up of 765 base pairs which indicates that modern human species lost around 37,000 genes during the split from its phylogenetic cousin, chimpanzee and other great apes (Gray, 7th August, 2015).

However, evolutionists and evolutionary biologists considered it a success as even after a loss of genes as well as genetic material, evolution of modern human species occurred from ancestral hominid and hominoid lineage which indirectly proved that a huge segment of human genome contained non-coding sequences. Scientists further suggested that shedding of unnecessary stretches of DNA or a number of non-coding sequences could have shaped the evolutionary progress of modern human species. In terms of evolutionary dynamics of any biological organism, if the loss of any structural entity or genetic constituent help the organism to survive in a changed environment, it indicates that this is a unique instance of reverse evolution. Hence, the evolutionary progress of life on Earth seems to be a gamble of nature where an apparent loss provides evolutionary success.

There is a subtle difference between "missing/losing" and "off-loading." For instance, the occurrence of the first happens as a part of ongoing process as a result of involuntary function. Whereas, "off-loading" is a voluntary action which is done consciously to initiate or execute a function. Hence, whenever geneticists and genome biologists have suggested loss of genetic information or entities from the genome and correlated it to evolution of modern human species, it has been presented as an instance of evolution of modern human species in the backdrop of "involuntary" loss of genes or genetic materials. Rather, it should be redefined as "off-loading" of genetic entities that get rid of excess or less important genetic material (non-functional DNA) to make the genome pro-active and expedite evolutionary functions to bring about the modern human species.

Every story has a beginning, a well-narrated ramification in the middle and a happy or tragic ending. Likewise, the unique story of loss of genetic materials from the human genome started with a respondent (Code named RP11) of an advertisement, published in a renowned US newspaper "The Buffalo News" in March, 1997, who agreed to donate 50 ml. of blood, as the main contributor along with 50 more volunteers. The blood samples were used for sequencing DNA to reconstruct the human genome, popularly referred to as the first effort to map the human genome, recognized as HGP or "Human genome project" (Zhang, 2018). The scientists, engaged in sequencing the DNA in HGP, propounded that 70% DNA material in the HGP was analyzed from the sample collected from RP11 and rest was analyzed with the samples of 50 volunteers. Hence, the mapping of the model human genome which was referred to as reference genome is the structurally configured genome of RP11. As each human individual contain a unique genetic code, there would always be subtle differences when a comparative genomic analysis is performed with the genome belonging to certain human communities (like representative of Latin American descent) with the reference genome. Hence, genomicists found that the comparative genomic analysis between the African descent communities and the reference genome showed that around 300 million letters of DNA were missing in the reference genome. Furthermore, a number of unique genes were discovered in the collective genome of 910 individuals of African descent that were absent in the reference genome (Sherman et al, 2018). The results in this comparative genomic analysis confirmed that Africans have a genetically richer genomic profile than any other racial community found in the world. It was proved that the scientists ignored these so-called "missing" or "evolutionary off-loading" of the stretches of DNA sequences as well as a substantial quantity of genes to evaluate its role in the occurrence of genetically inherited diseases.

Rachel Sherman, a eminent computational biologist, along with her fellow researchers, under the mentorship of Steven Salzberg at Johns Hopkins University, was involved in determining the genetics of asthma in the people of African descent (Zhang, 2018). In order to carry out the comparative genomic analysis, Sherman et al (2019) used intensively sequenced datasets of the 910 individuals of African descent (living in Caribbean, West Africa, South America and North America) to form the absent DNA sequences in the reference genome and was able to sequence 1.19 trillion sequence to form the African model genome. Sherman and her colleagues revealed that the African model genome contained 10% more DNA with 315 unique protein encoding genes in it than the reference genome constructed earlier (Sherman et al, 2019). Such observation created an opportunity to study the qualitative assessment of DNA missing from the reference genome. In 2019, when Sherman and Salzberg finally revealed that 300 million letters were missing in 125,715 segments of DNA, it motivated molecular geneticists to explore those missing letters that could contain unique coding potential integrated to mutation-related disease (Sherman et al, 2019). However, Tina Graves-Lindsay, another contemporary, famous geneticist of Washington University, USA, pointed out another limitation in the genomic model of African descent being compared to the reference genome, RP11, as she considered the erstwhile reference genome model (GRCh38) in HGP project to be of African-American descent, which might not help to find the missing pieces of DNA letters. Such limitation should never be ignored when a reference genome is determined from any individual of any particular geographical location.

In order to sort through the biased versions of reference genome in the human species, Graves-Lindsay (Zhang, 2018) along with her research colleagues, sequenced 15 sample (5 samples of which were of African descent), representing diversified DNA constituents possessed by human communities. Still, such an endeavor by Graves-Lindsay was not free from error as every time the scientists tried to upgrade the experimental protocol to get better result some flaws remained which failed to accommodate the sequences of distant Australian and Polynesian communities and left a big chunk of missing DNA from their reference model (Zhang, 2018). According to Deanna Church, the founding member of Genome Research Consortium, the genomicists did not realize the importance of genetic diversity of the modern human species, so cascading mutational cycles of deletion, insertion of bases rendered a number of letters missing and made the genome complicated to compare with any reference genome unless a thorough assessment was done on these "missed" or "offloaded" bases first.

It did not underestimate the first effort of HPG to form the reference genome with RP11 along with 50 volunteers.

Along with a series of missing segments of DNA and genes, factors like pseudogenation or mutational transformation of protein encoding genes to non-functional genes was found to be one of the important factors for evolutionary change to help the emergence of human species (Wang et al, 2006). In order to find out whether the process of pseudogenation were "selectively favored" to drive human evolution or not, Wang et al (2006), conducted a composite study of comparative genomic analysis and secondary literature survey to trace around 80 non-activated pseudogenes in the human lineage which changed from active to non-active after the human lineage diverged from the chimpanzee. Though there is no remarkable difference between the human genome and chimpanzee genome in terms of genomic sequence and protein sequence, genomicists observed that the major biological differences between these two groups are marked by features like brain size, bipedalism, language communication and susceptibility to HIV (Human Immuno-deficiency Virus) (Chen and Li, 2001). Intensive studies in the field of molecular genetics and comparative genetics revealed that the functional role of the transcription factor FOXP2 in the human genome was integrated into the language-mediated communication ability of human species and involved two adaptive amino acid replacements in evolution of the hominid lineage (Lai et al, 2001).

It was further witnessed that a number of amino-acid replacements and change in protein sequence occurred during the evolution of human species. Likewise, the loss of a number of functional genes trans-mutated to non-functional form have been ascertained in hominin evolution by means of pseudogenation in the post-human-chimpanzee lineage divergence (Szabo et al, 1999; Winter et al, 2001; Stedman et al, 2004; Perry et al, 2006). A contemporary hypothesis on evolutionary functions by Olson (1999) and Olson and Varki (2003) proposed their "less is more" hypothesis where loss of genes is the primary force that put the evolutionary process on a roll. This "less is more" hypothesis by Olson and Varki (2003) intrigued genomicists and molecular biologists to envision that loss of a number of genes and pseudogenation such as, the inactivation of gene, encoding CMAH (CMP-N-acetylneuraminic acid hydroxylase), mediated deficiency of N-glycolylneuraminic acid on the cellular surface of human skin was triggered by the replacement of Alu sequence around 2.7 million years ago

(Hayakawa et al, 2001). Furthermore, the act of pseudogenation of sarcomere myosin gene masticatory myosin heavy chain 16 (MYH16) resulted in the loss of erstwhile masticatory muscles but expedited the increasing size of the human brain (Perry et al, 2006) and helped the hominid lineage to evolve and adapt to the penultimate level of primate evolution.

In the recent works of Wang et al (2006), scientists assessed the evolution-related gene losses of human lineage by comparing the "non-processed pseudogenes" with the chimpanzee genome and these pseudogenes were found to have evolved around 6-7 million years ago after the phylogenetic split of human from chimpanzee lineage (Brunet et al, 2002). In order to examine whether pseudogenation is adaptive or not, Wang et al (2006) conducted a Cysteinyl aspartate proteinase (CASPASE12) mediated experiment on "inflammatory and immune response to endotoxins." The scientists noticed that the fixation of a null allele[13] at CASPASE12 was closely integrated to positive selection as they assumed that it stalled the chance of sepsis; though its selective advantage was found to be around 0.9% (Wang et al, 2006). The researchers further assumed that such adaptive gene losses might have been induced by environmental changes. They revealed that the act of pseudogenation which has driven the evolution of modern human species, supposedly, took place during migration of the earliest members of modern human species out of Africa and the subsequent dispersal all over the world (Wang et al, 2006). So, the "less is more" hypothesis of Olson (1999) was validate. It elucidate the potential evolutionary function of non-functional DNA as well as the treasure-trove of genes in the genome storage. The beneficial loss of such genes or "evolutionary off-loading of heavily-loaded genetic material" could also help the biological evolution in moving forward.

However, it needs to be noted that extraordinary, technological advancement in the field of genomics and biotechnology, specifically, exceptional level of progress in the field of DNA sequencing, PCR amplification, utilization of restriction enzymes, cloning etc. helped scientists use DNA technology in solving multifarious issues ranging from DNA bar-coding to forensic analysis to determination of evolutionary

---

[13] Null Allele: The trans-mutated gene or non-functional copy of a gene which does not have any power to yield any gene product as well as protein encoding potential that has not been observed in phenotypic level.

relationships to solving phylogenetic problems. So, in order to solve any DNA technology-depended issues, precise collection protocol of DNA samples needs to be followed to ensure its quality and quantity which would facilitate flawless execution of sequencing and/or cloning of DNA to get error-free result. Thus, damages of DNA sample at either the time of extraction (improper extraction methodologies) or the pre-collection stage (due to improper preservation under adverse exposition to external factors in natural environment) was found to be a major issue leading the scientists to discard the samples for analysis. However, in a number of instance DNA damages were mainly due to erroneous replication, catalyzed by the enzyme DNA polymerase (Lindahl and Wood, 1999), thus the type of DNA damages have been categorized under three categories:

A. Hydrolysis of N-glycosyl bond: According to molecular geneticists, the N-glycosyl bond is the most unstable bond in the DNA molecule which holds the base at one end and the deoxyribose sugar back bone at the other. The hydrolysis of N-glycosyl bond expel the base contents of the DNA molecule to render a apurinic/apyrimidinic site (AP site), leading to the formation of a nick while adjacent to freeze-dried water molecules. Genomicists noticed that in human genome around 2000-10,000 AP sites were formed in a 24-hour cycle (Lindahl, 1993). They also noticed that AP sites in the DNA stalled formation of base pairing with the synthesized nucleotide during DNA replication which stopped the working of PCR polymerase enzyme disrupting the amplification and sequencing process (Sikorsky et al, 2007). The fragmentation of DNA molecules were found to occur at the nick-end of AP sites expediting the chance of missing genetic materials. (Evans, Jr., 2007)

B. Hydrolytic deamination of cytosine: Like AP site-induced damage on DNA molecule as well as missing gene, deamination of cytosine to uracil (which is the constituent base of RNA) in the DNA molecule was considered a major physiological damaging process of DNA (Lindahl, 1993). The sequencing of ancient DNA was, technically, a cumbersome process due to structural alteration of DNA into cytosine deamination (Stiller et al, 2006; Gilbert, 2007). In this hydrolytic damage of the DNA molecule, the single-stranded

DNA is more prone to cytosine deamination-induced damage than double-stranded DNA (Evans, Jr., 2007). Sscientistes observe that the active site of different polymerases constitute a binding site for deaminated cytosine and the involvement of deaminated cytosine in DNA sequencing and amplification was found to be polymerase dependent (Fogg et al, 2002). It was experimentally found that some DNA such as, Taq polymerase DNA, induce generation of mutated DNA strand in the process of replication which might cross the deaminated cytosine (uracil). Instead of inserting the complementary base, Guanine, this polymerase inserts Adenine residue which is perfectly complementary to uracil (Evans. Jr., 2007). The experimental observations of molecular geneticists like Lasken et al (1996) ascertained that archaeal polymerases such as, Pfu polymerase, Vent polymerase, etc., stopped the polymerization of DNA strand as it reached dexoyuracil and this automatic shut-off process of polymerization helped to prevent synthesis of mutagenic DNA strand (Lasken et al, 1996; Evans, Jr., 2007). So, the inhibitory action of the progress of polymerization and PCR of deaminated cytosine has an inherent potential to stop mutagenesis of long stretches of DNA sequences (Evans, Jr., 2007).

C. Hydrolytic damage due to oxidation: Scientists found that the most common type of DNA damage or decay is oxidation-induced damage. As a result of long storage in natural environment, long-term exposition in oxygen or prolonged oxidation of bases render its structural configuration weak. Observations of the base pairing experiments by genomicists revealed that oxidation modified alteration of the purine-derivative Guanine to 8-oxo-guanine creates base incompatibility (Lindahl, 1993). Normally, the purine-derivative Guanine is paired with the pyrimidine-derivative Cytosine, however, scientists observed that the modified form of Guanine—8-oxo-guanine—is paired with the pyrimidine-derivative Adenine instead resulting in synthesis of mutagenic DNA in mitochondria (Evans Jr., 2007). Scientists observed that oxidative damage is one of the most common type of damage—fragmentation and potential loss of genetic materials—as it was found to trigger the ageing process in the human species (Weissman et al, 2006).

Often, issues like exposure of DNA molecule under UV radiation would lead to mutagenic DNA by altering the function of DNA polymerase and damage or fragment DNA molecule (Evans, Jr., 2007). The DNA-DNA cross-linking or formalin-induced cross-linking of DNA plays an important role in DNA damage or fragmentation or loss of important genes by stalling the progress of polymerization and denaturation of DNA (Evans, Jr., 2007). As the pH of the formalin-mixed solution decelerated with the passage of time which resulted in the formation of formic acid, enhancing the AP sites and causing fragmented stretches of DNA and, ultimately, loss of genes in the long run (Kelly, 2006).

Though in the backdrop of a dynamic environment, a number of changes in genetic material like DNA damage, loss of gene and formation of mutagenic DNA, was found to be the common genetical function and integral part of evolutionary function witnessed in diverse groups of organisms. It was also witnessed that the usual and intermittent extent of damages in DNA molecule or loss of genetic material would not make any big changes in the genetic profile of the organism or evolve as a new variety of the existing lineage or a new species. So, it indirectly confirmed that the genome of the concerned organism must act as a super storage of excess DNA and possibly contain more than one copy of all essential genes so that a certain level of damage of the extant DNA threads or loss of few genes would not alter the genetic profile or turn positive selection to negative ones. Hence, frequent emergence of newly-evolved organisms is considered to be a chance factor and the continuation of the existing lineage of species is the ultimate reality in the course of evolution. (Kundu, 2018)

However, evolutionary biologists have assumed for a long time that sudden and irreversible changes of DNA in human genome or mutations have played a key role in the evolution of a number of unique traits. Evolutionary biologists further ascertained that the ultimate role of these newly-evolved unique traits drive out mutational changes, depending upon its wide-spread ramification across the population of the concerned organism. They noticed that though every individual possess 2 copies of each type of genes, (alleles) the copies vary at inter- and intra-community level of the present human species. Population geneticists observed that mutational changes in one copy of the gene might change aspects, like resistance against HIV, than another copy, like having no or less resistant against immune-deficiency. If this newly-evolved, hypothetical HIV resistance favored selection, the individuals with HIV resistance potential would be able to generate more number of offsprings, thereby more copies of such gene variants encoding HIV resistance, and would witness widespread circulation in following generations throughout the population (Pennisi, 2016). Referring to two contemporary case studies—the gradual emergence of taller and fairer communities of British descent since the Roman era and the gradual dwindling of smoking cigarette in certain communities—genome and evolutionary biologists have critically reviewed the "genetic shift" at the population level which is the impetus for driving "evolution in action" and is integrated to change in genomes over a period of time varing from few thousand to millions of years (Pennisi, 2016). Molly Przeworski, a popular evolutionary biologist from Columbia University, USA, hypothesized that the structural conformation of the genome (which constitute high content of non-coding DNA molecules) seems to be evolutionary proactive in responding to subtle forms of genetic change in its gene. She stated that "being able to look at selection in action is exciting" and that "it's a game changes in terms of understanding evolution." (Pennisi, 2016)

So, evolutionary scientists used unique molecular data, like allele frequency of a gene variant, driven by single-base change, which is referred to as singleton, to monitor the evolutionary shifts of any biological organism (Pennisi, 2016). Jonathan Pritchard, another eminent genomicist of Stanford University, USA, and his research fellows have developed a scoring pattern called "Singleton Density Score" or SDS (based on the density of nearby single mutations) while working on research investigations pertaining to determination of allele frequency of 3000 genomes called "UK10K" sequencing project (Pennisi, 2016). The scientists concluded that "the more intense the selection on allele, the faster it spreads, and the less time there is for singletons to accumulate near it" (Pennisi, 2016). Following the footprints of Pritchard, a number of researchers critically reviewed the SDS results of the UK10K project and suggested that about 2000-3000 years ago, the ancestral lineage of modern Britons went through some evolutionary changes. Strong selection triggered by allele frequency of a single gene, across the genome variant resulting in modern Britons evolving with taller stature, blond hair, fair complexion and blue eyes (Field et al, 2016). It helped the scientists further to assume that genes for lighter hair and eye color was also responsible for fair skin color which expedite the production of Vitamin D in low sunlight environment (Pennisi, 2016). It helped the scientists to assume that polygenic adaptation was responsible for regulating the phenotypic and genotypic variation in the diversification of modern human species (Field et al, 2016).

In his experiment on allele frequency, Pritchard and his colleagues, have also noticed that it's not a single gene but variation of a number of genes that controlled the selection of the trait. As an example, they stated height, head, circumference of the infant and hip size of the female which is an important feature for giving birth, was regulated by 4 million variant DNAs in the genome (Field et al, 2016). The role of genetic changes in the evolutionary progress of human species was further advanced by Joseph Pickrell, a great evolutionary geneticist of New York Genome Center, USA, who found that "late-onset of menstruation" in women helped them to live longer as it was assumed that it delayed the ageing process and death. Pickrell's work revealed that the decreasing number of "ApoE4" allele, responsible for causing Alzheimer's disease in older people was considered to be the cause of early death for its carrier and has indicated that selection plays a role in shaping evolution of modern human species to a certain extent in the backdrop of gene-environment interactions (Pennisi, 2016). Very recently, evolutionary geneticists noticed that during the loss of DNA from non-coding regions of the genome, introns are least affected indicating that selective pressure would not involve introns as part of the evolutionary progress (Rigau et al, 2019). All evolutionary experimental process could be tested in the dynamic background of natural environment when the genome tool-box is fully equipped to deal with any usual or unusual situations concerning the survival of the organism. The Junk or non-coding DNA of the human genome plays a cardinal role to ensure living and continue the journey of human evolution.

References

Brunet, M., Guy, F., Pilbeam, D., Mackaye, H.T., Likius, A., Ahounta, D., Beauvilain, A., Blondel, C., Bocherens, H., Boisserie, J.R., De Bonis, L., Coppens, Y., Dejax, J., Denys, C., Duringer, P., Eisenmann, V., Fanone, G., Fronty, P., Geraads, D., Lehmann, T., Lihoreau, F., Louchart, A., Mahamat, A., Merceron, G., Mouchelin, G., Otero, O., Pelaez Campomanes, P., Ponce De Leon, M., Rage, J.C., Sapanet, M., Schuster, M., Sudre, J., Tassy, P., Valentin, X., Vignaud, P., Viriot, L., Zazzo, A., and Zollikofer, C. (2002). A new hominid from the Upper Miocene of Chad, Central Africa. Nature. 418(6894): 145-51. doi: 10.1038/nature 00879. Erratum in: Nature 2002 Aug 15;418(6899):801.

Chen, F.C., and Li, W.H. (2001). Genomic divergences between humans and other hominoids and the effective population size of the common ancestor of humans and chimpanzees. Am J Hum Genet. 68(2):444-56. doi:10.1086 /318206.

Evans Jr., T.C. (2007). DNA Damage - the major cause of missing pieces from the DNA puzzle. https://international.neb.com/tools-and-resources/ feature-articles/dna-damage-the-major-cause-of-missing-pieces-from-the- dna-puzzle

Field, Y., Boyle, E.A., Telis, N., Gao, Z., Gaulton, K. J., Golan, D., Yengo, L., Rocheleau, G., Froguel, P., McCarthy, M.I., and Pritchard, J.K. (2016). Detection of human adaptation during the past 2000 years. Science. 354(6313):760-764. doi: 10.1126/science.aag0776. Epub 2016 Oct 13.

Fogg, M. J., Pearl, L.H., and Connolly, B.A. (2002). Structural basis for uracil recognition by archaeal family B DNA polymerases. Nat Struct Biol. 9(12):922-7. doi: 10.1038/nsb867.

Gilbert, M.T., Binladen, J., Miller, W., Wiuf, C., Willerslev, E., Poinar, H., Carlson, J.E., Leebens-Mack, J. H., and Schuster, S.C. (2007). Recharacterization of ancient DNA miscoding lesions: insights in the era of sequencing-by-synthesis. Nucleic Acids Res. 35(1):1-10. doi: 10.1093/nar/gkl483.

Gray, R. (7th August, 2015). Humans LOST DNA as they evolved: Early species had the equivalent of thousands more genes than we do now. https://www. dailymail.co.uk/sciencetech/article-3187857/Humans-LOST-DNA- evolved-Early-species-equivalent-thousands-genes-now.html

Harper, J.M., Salmon, A. B., Leiser, S. F., Galecki, A. T., and Miller, R. A. (2007). Skin-derived fibroblasts from long-lived species are resistant to some, but not all, lethal stresses and to the mitochondrial inhibitor rotenone. Aging Cell.6(1): 1-13. doi: 10.1111/j.1474-9726.2006.00255.x. Epub 2006 Dec 5.

Harper, P. S. (2006). The discovery of the human chromosome number in Lund, 1955-1956. Hum Genet.119(1-2):226-32. doi: 10.1007/s00439-005-0121- x. PMID: 16463025.

Hayakawa, T., Satta, Y., Gagneux, P., Varki, A., and Takahata, N. (2001). Alu-mediated inactivation of the human CMP- N-acetylneuraminic acid hydroxylase gene. PNAS U S A. 98(20):11399-404. doi: 10.1073/pnas.191268198.

IJdo, J.W., Baldini, A., Ward, D.C., Reeders, S. T., Wells, R. A. (1991). Origin of human chromosome 2: an ancestral telomere-telomere fusion. PNAS U SA. 88(20): 9051-5. doi: 10.1073/pnas.88.20.9051.

Kelly, K. (2006). Path to Effective Recovering of DNA from Formalin-Fixed Samples in Natural History Collection, Washington, DC: The National Academies Press. (pp. 5–14).

Kundu, S.R. (2018). Origins and Evolution of Plants on the Earth and the Descendants of ANITA . Nova Science Publishers Inc. p.387.

Lai, C.S., Fisher, S.E., Hurst, J.A., Vargha-Khadem, F., and Monaco, A.P. (2001). A forkhead-domain gene is mutated in a severe speech and language disorder. Nature. 413(6855): 519-23. doi: 10.1038/35097076.

Lasken, R.S., Schuster, D.M., and Rashtchian, A. (1996). Archaebacterial DNA polymerases tightly bind uracil-containing DNA. J Biol Chem 271(30): 17692–17696.

Lindahl, T. (1993). Instability and decay of the primary structure of DNA. *Nature* 362: 709–715.
https://doi.org/10.1038/362709a0

Lindahl, T., and Wood, R. D. (1999). Quality control by DNA repair. Science. 286(5446):1897-905. doi: 10.1126/science.286.5446.1897.

Meyer, M., Kircher, M., Gansauge, M.T., Li, H., Racimo, F., Mallick, S., Schraiber, J.G., Jay, F., Prüfer, K., de Filippo, C., Sudmant, P.H., Alkan, C., Fu, Q., Do, R., Rohland, N., Tandon, A., Siebauer, M., Green, R.E., Bryc, K., Briggs, A.W., Stenzel, U., Dabney, J., Shendure, J., Kitzman, J., Hammer, M.F., Shunkov, M. V., Derevianko, A.P., Patterson, N., Andrés, A. M., Eichler, E.E., Slatkin, M., Reich, D., Kelso, J., Pääbo, S. (2012). A high-coverage genome sequence from an archaic Denisovan individual. Science. 338(6104):c222-6. doi: 10.1126/science.1224344.

Olson, M. V. (1999). MOLECULAR EVOLUTION '99 When Less Is More: Gene Loss as an Engine of Evolutionary Change. Am. J.Hum. Genet. 64: 18-23.

Olson, M. V., and Varki, A. (2003). Sequencing the chimpanzee genome: insights into human evolution and disease. Nat Rev Genet. 4(1):20-8. doi: 10.1038/nrg981.

Pennisi E. (2016). HUMAN EVOLUTION:Field Tracking how humans evolve in real time. Science 352(6288):876-7. doi: 10.1126/science.352.6288.876.

Perry, G. H., Tchinda, J., McGrath, S. D., Zhang, J., Picker, S. R., Cáceres, A.M., Iafrate, A.J., Tyler-Smith, C., Scherer, S.W., Eichler, E.E., Stone, A.C., and Lee, C. (2006). Hotspots for copy number variation in chimpanzees and humans. PNAS U S A. 23;103(21): 8006-11. doi: 10.1073/pnas.0602318103.

Rejon, M. R. (17th January, 2017). The Origin of the Human Species: a Chromosome Fusion?
https://www.bbvaopenmind.com/en/science/ bioscience/the-origin-of-the-human-species-a-chromosome-fusion/

Rigau, M., Juan, D., Valencia, A., and Rico, D. (2019). Intronic CNVs and gene expression variation in human populations. PLoS Genet 15(1): e1007902. https://doi.org/10.1371/journal.pgen.1007902

Sherman, R.M., Forman, J., Antonescu, V., Puiu, D., Daya, M., Rafaels, N., Boorgula, M. P., Chavan, S., Vergara, C., Ortega, V.E., Levin, A.M., Eng, C., Yazdanbakhsh, M., Wilson, J.G., Marrugo, J., Lange, L.A., Williams, L.K., Watson, H., Ware, L.B., Olopade, C.O., Olopade, O., Oliveira, R.R., Ober, C., Nicolae, D. L., Meyers, D.A., Mayorga, A., Knight-Madden, J., Hartert, T., Hansel, N.N., Foreman, M.G., Ford, J.G., Faruque, M.U., Dunston, G.M., Caraballo, L., Burchard, E.G., Bleecker, E.R., Araujo, M.I., Herrera-Paz, E.F., Campbell, M., Foster, C., Taub, M.A., Beaty, T.H., Ruczinski, I., Mathias, R.A., Barnes, K.C., Salzberg, S.L. (2019). Assembly of a pan-genome from deep sequencing of 910 humans of African descent. Nat Genet. 51(1):30-35. doi: 10.1038/s41588-018-0273- y. Epub 2018 Nov 19. Erratum in: Nat Genet. 2019 Feb;51(2):364.

Sikorsky, J.A., Primerano, D.A., Fenger, T.W., and Denvir, J. (2007). DNA damage reduces Taq DNA polymerase fidelity and PCR amplification efficiency. Biochem Biophys Res Commun. 355(2):431-7. doi: 10.1016/j.bbrc.2007.01.169.

Stedman, H.H., Kozyak, B.W., Nelson, A., Thesier, D.M., Su, L.T., Low, D.W., Bridges, C.R., Shrager, J.B., Minugh-Purvis, N., and Mitchell, M. A. (2004). Myosin gene mutation correlates with anatomical changes in the human lineage. Nature. 428(6981):415-8. doi: 10.1038/nature02358.

Stiller, M., Green, R.E., Ronan, M., Simons, J.F., Du. L., He, W., Egholm, M., Rothberg, J.M., Keates, S.G., Ovodov, N.D., Antipina, E.E., Baryshnikov, G.F., Kuzmin, Y.V., Vasilevski, A.A., Wuenschell, G.E., Termini, J., Hofreiter, M., Jaenicke-Després, V., and Pääbo, S. (2006). Patterns of nucleotide misincorporations during enzymatic amplification and direct large-scale sequencing of ancient DNA. PNAS U S A. 103(37):13578-84. doi: 10.1073/pnas.0605327103. Epub 2006 Aug 25. Erratum in: Proc Natl Acad Sci U S A. 2006 Oct 3;103(40):14977.

Szabó T.A., (1999). Genetic erosion, human environment and ethnobiodiversity studies. *Preprint in:* Bio Tár Electronic. Germoplasma *BTN 766: 1-16.* Bge766ba99050604 Praga, W:
http://genetics.bdtf.hu and http://vebi. vein.hu

Tjio, J. and Levan, A. (1956). The chromosome number of Man. Hereditas, 42(1-2): 1-6.

Wang, D., Amornsiripanitch, N., and Dong, X. (2006). A Genomic Approach to Identify Regulatory Nodes in the Transcriptional Network of Systemic Acquired Resistance in Plants. PLoS Pathog 2(11): e123. https://doi.org/ 10.1371/journal.ppat.0020123

Weissman, D. H., Roberts, K. C., Visscher, K. M., and Woldorff, M. G. The neural bases of momentary lapses in attention. Nat Neurosci. 9(7):971-8. doi: 10.1038/nn1727.

Winter, H., Langbein, L., Krawczak, M., Cooper, D. N., Jave-Suarez, L.F., Rogers, M. A., Praetzel, S., Heidt, P.J., and Schweizer, J. (2001). Human type I hair keratin pseudogene phihHaA has functional orthologs in the chimpanzee and gorilla: evidence for recent inactivation of the human gene after the Pan-Homo divergence. Hum Genet. 108(1): 37-42. doi: 10.1007/s004390000439.

Zhang, W., Cai, X., Yu, J., Lu, X., Qian, Q., Qian, W. (2018). Exosome-mediated transfer of lncRNA RP11-838N2.4 promotes erlotinib resistance in non-small cell lung cancer. Int J Oncol. 53(2):527-538. doi: 10.3892 /ijo.2018.4412. Epub 2018 May 17.

Conclusion

Basis of Race and Its Integration in Human Evolution: In the Backdrop of "DNA Discussion Project"

All these stories have been suppressed pop out in genes...I grew up in the 1960s when light skin was really a big deal. So, I think of myself as being pretty brown skinned. I was surprised that a quarter of my background was European. It really brought home this idea that we make race up...That race is a human construction doesn't mean that we don't fall into different groups or there's no variation. But if we made racial categories up, maybe we can make new categories up, maybe we can make new categories that function better.

---- Anita Foreman

There is a popular saying that "every story has an end, but in life, every ending is just a new beginning." For us, the modern human species (*Homo sapiens*), it's really important to know our past, our history of origin. And after a long time, scientists have finally ascertained that around 300,000 years ago we emerged with our unique features somewhere in Africa where we stayed for the next 200,000 years before migrating out of Africa. This migration triggered a trail of diversification which led to interactive progress of genetic changes (like cascade of mutations, selection pressure) and phenotypic distinctions in the backdrop of dynamic environments and finally diversified in the form of distinct "racial structures" which ushered a new era in the socio-biological history of our evolution. Ideas and interpretation of "racial structures" has gone through a labyrinth of conflicts such whether there is any scientific basis to race or is it merely a "Made-up label." Skipping this part of journey of human evolution, means an incomplete understanding of the genetic basis of human evolution.

The first studies on racial structure was started by Doctor Samuel Morton in USA, in the 19th century, who was famous for collecting and scientifically studying skull architecture to review the anatomical features of braincase of a wide array of human skulls (either collected from battlefields or graveyards) and segregate those skulls into certain racial categories (following a certain hierarchy) on the basis of comparative similarities or dissimilarities. On the basis of his self-developed skull examining protocol (which is called craniometrics), Samuel Morton primarily categorized the entire human species into five distinct races (Kolbert, 12th March, 2018). Though molecular anthropologists later found that Morton's way of classification of human race adhered to predisposed ideas where he considered that White or "Caucasians" (European descent) were the most intelligent, followed by hierarchical level of intelligence in "Mongolians" (East Asian descent), "Southeast Asians" then by "Native Americans" and lastly by Blacks or "Ethiopians" (Kolbert, 12th March, 2018). Paul Mitchell, a renowned anthropologists and a faculty member of University of Pennsylvania observed that in the United States Civil War, the southern states had used Morton's ideas to promote slavery and ensure the supply of black African people from Dutch and Portuguese slave traders (Kolbert, 12th March, 2018). After the death of Samuel Morton in 1851, an obituary was published in the "Charleston Medical Journal," California, USA, where the following derogatory comment was printed:

> "*Giving to the negro his true position as an inferior race.*" (Kolbert, 12th March, 2018) So, the reality of the racial distinctions of modern human species did not comply with the objectivity of science. Rather, Morton's academic attributes on "*racial structures*" was used for centuries to achieve socio-political gain which has influenced human history and has been a witness to conflicts and dominance between racial hierarchy since 18th century.

Samuel Morton's work, which was mainly based on "immutable and inherited differences among people," was published little earlier than Charles Darwin theory of evolution and existed much earlier than the concept of DNA as genetic material. So, at the end of Human Genome Project (HGP) in June, 2000, when Craig Venter, the renowned genome biologist, presented the key observations of HGP project in a ceremony at White House and stated that "the concept of race has no genetic and scientific basis" it jolted century-long perception of racial distinctions (Kolbert, 12th March, 2018).

In 2006, Anita Foreman, an eminent professor of communications in

West Chester University, Pennsylvania, USA, started a programs known as "DNA Discussion Project," which intended to gather and review the ancestry of individuals and determine their family history in terms of the lineage of its ethnicity, collected via personal communication and to substantiate it with the genetic information of their past analyzed from their DNA samples collected from their saliva. (Kolbert, 12th March, 2018) She found that the individual's perception of their own ethnicity did not necessarily substantiate their genealogical origin encrypted in the DNA. She referred to a few case studies which are as follows:

   a. A woman believing that she did not have a Middle Eastern ethnic background.

   b. A young woman, having a family origin of Indian ethnicity, was found to be of Irish rootstock in reality.

   c. A woman who believed her grandparents to be of Native American origin, was found to have no such ancestral history.

   d. A young man having biracial features was found to have solely come from the European rootstock.

   e. An individual, raised as an orthodox Christian, found that their ancestral rootstock was of Jewish origin.

These instances made Foreman state, "All these stories that have been suppressed pop out in the genes." She recounted her own experience stating, "I grew up in 1960s when light skin was really a big deal...so, I think of myself as being pretty brown skinned." (Kolbert, 12th March, 2018) As the coordinator of the DNA Discussion Project, Foreman, herself identified as African-American but her DNA test results revealed that her ancestral attributes were inherited partly from Ghana (Africa) and partly from Scandinavia (Europe) which made her comment on her own ethnicity stating:

> I was surprised that a quarter of my background was European.....It really brought home this idea that we make up race. (Kolbert, 12th March, 2018).

As a matter of fact the human skin color, which is regulated by melanin content has been expressed as a cumulative interaction of a number of genes. Evolutionary biologists hypothesized as time elapsed the cascades of mutational cycle altered the configurations of genes and interacted with each other to emerge in distinct races with diverse skin color as our ancient predecessors left behind their place of origin, Africa, to settle down in different geographical locations in different climatic, altitudinal regions across the biosphere. The evolutionary biologists contemplated further that "genetic changes are the result of random mutations—tiny tweaks to DNA, the code of life. Mutations occur at a constant rate, so the longer a group persists in handing down its genes, the more tweaks these genes will accumulate." (Kolbert, 12th March, 2018) After reviewing the evolutionary history of human species, molecular anthropologists stated that the earliest record of a family tree—Khoe-San—was found to originate in South Africa, whereas, the Pygmies evolved from Central Africa which affirms the "Out of Africa" hypothesis of the origin of modern human species. It helped molecular anthropologists to redefine racial distinctions as diversifications of the African sub-population like Pygmies, Khoe-San as they took the stride to disperse out of Africa and colonize diverse environmental condition in diverse latitudes, rather than the stereotypic racial diversification dividing the human population into Caucasians, Blacks, Native Americans, and Asians (Kolbert, 12th March, 2018).

The history of modern human species has gone through cascades of evolution as anthropologists tried to decrypt the genetic codes of modern human species and found that around 60,000 years ago, the ancestors of the modern human species migrated from Africa to explore, Eurasia, Middle-East Asian regions, etc., where they interbred with Neanderthals and Denisovans and gave birth to the immediate ancestral lineage of the modern human species (Kolbert, 12th March, 2018). On the basis of molecular anthropological evidence, as well as the composite studies of decrypting results of modern human genome coupled with paleontological evidence, it helped scientists trace the earliest history of human migration related to the evolution of modern human species. Scientists further stated that some of the earliest migrants settled in Australia around 50,000 years ago, while some others in Siberia around 45,000 years ago. Some migrated to South America around 15,000 years ago and mutational changes in their genetic elements helped them to survive in the distinct environmental conditions which resulted in a wide array of skin and hair colors and distinct shaped and sized features (Kolbert, 12th March, 2018). Evolutionary biologists noticed that mutational changes of human genetic materials were either harmful or neither harmful nor beneficial. In some situations, mutational changes of genetic materials were found to be beneficial. If and when such mutational changes were assisted by natural selection to spread throughout the entire population, it rendered the evolutionary functions work. For example, some of the human sub-populations at high altitudinal regions in like Andean Atilpano, Ethiopian Highland, Siberian Traps, Tibetan plateau, etc. have specially adapted to survive in adverse environmental conditions with low oxygen levels which were caused by mutational changes in their genetic materials to ensure their survival (Kolbert, 12th March, 2018).

In 2013, Yanna Kamberov, a geneticist in Perelman School of Medicine, University of Pennsylvania, USA, along with her co-researchers, experimented on mouse model with the adaptive gene variant of human called *EDARV370*, an ectodysplasin receptor found in the genome of Native Americans and East Asians but rarely found in the genomes of Europeans and Africans (Kamberov et al, 2013). They observed that presence of *EDARV370* variant in East Asians and Native Americans led to the sub-populations having thick brown hair on the outer skin of the body, thicker hair, presence of excessive number of sweat glands and little extra fat in the mammary glands (Kamberov et al, 2013). The computational modelling helped Kamberov and her associates to confirm the evolution of this allele around 30,000 years ago in some part of Central China (Kamberov et al, 2013). Though the impact of the mutational changes of the genetic materials of *EDARV370* needs to be further investigated this experimental model gives a preliminary idea on the generation of adaptive variation in the evolution of modern human species. By referring to the basic constituents of the human genome, in her article titled *"There's No Scientific Basis for Race-it's a Made-Up Label,"* Elizabeth Kolbert (12th March, 2018) stated:

> *The human genome consists of three billion base pairs—page after page of A's, C's, G's and T's—divided into 20,000 genes. The tweak that gives East Asians thicker hair is a single base change in a single gene, from T to a C.*

In 2008, a similar type of experiment was carried out by Keith Cheng, a popular pathologist and geneticist in Penn State College of Medicine, USA, who developed a Zebra-fish model where he tried to find the impact of mutational changes in genetic material by regulating the vertebrate's pigmentation. Cheng noticed that he was able to breed zebra fish with lighter stripes regulated by mutation of pigment gene (*SLC24A5*), which is also present in Europeans with lighter skin color. The pigment-forming gene *SLC24A5* constituted 20,000 base pairs and is almost structurally identical across the modern human species with the little difference of one base location of the nucleotide where *G*'s were found in sub-Saharan African gene variant and *A*'s in Europeans (Cheng, 2008). The comparative genomic analysis between sub-Saharan Africans and the Europeans by contemporary molecular anthropoligists along with substantial review of their migrational histories and cultural anthropological evidence revealed that mutational change of the *SLC24A5* variant from G's to A's supposedly evolved around 8,000 years ago (Cheng, 2008).

Going one step further, Heather Norton, an eminent molecular anthropologist in University of Cincinnati, USA, denounced the earlier unscientific distinction for 'human race' and she firmly stated, "I think it can be very powerful to explain to people that these changes we see, it's just because I have an *A* in my genome and she has a *G*." (Kolbert, 12th March, 2018). As a matter of fact, for better understanding about ourselves, "Who We Are and How We Got Here," the concern of David Reich, one of the famous paleogeneticist, of Harvard University, USA, is extremely important where he stated:

What the genetics show is that mixture and displacement have happened again and again and that our pictures of past 'racial structures' are almost wrong. (Kolbert, 12th March, 2018).

In 2006, Anita Foreman's "DNA Discussion Project' created an atmosphere where participants would have better opportunity to know about their ethnicity, going above and beyond the stereotypic perception of racial identity like "Japanese," "Chinese," "American-Indian" etc. The "DNA Discussion Project" created an open environment to understand not only the facts and findings of our biological evolution but also an opportunity to recognize our socio-cultural evolution started since the mass migration of our ancestors out of Africa. Hence, Anita Foreman's endeavor is undoubtedly a self-identifying orientation where as a descendant of modern human species, we recognize, understand and share our stories about our ancestors, our past who went through a long but unavoidable tangle to emerge as modern human species.

Like every story has a beginning, a middle and an end, the evolutionary journey of life has had the same template of progress which chronicles the history of origin and diversification of any organism, specifically, ourselves as modern human species. Therefore, evolutionary journey is not only the struggling narratives of our past or a glorified journey to our future but also an acknowledgement of our identity, origin, and evolution as a species on the Earth. Our progress is to understand our ecological role in this biosphere at inter- and intra-specific level structurally and functionally. All these answers are locked in our genome and vast swathes of it still remains shrouded in darkness. Truth is an ideal example of abstract noun and which neither glorifies nor maligns any facts, findings, belief or perceptions. Its cardinal role is just proper presentation of the concerned entity and it has the power to expose the concerned entity's identity and function. The cumulative endeavor by evolutionary biologists, molecular biologists, biotechnologists, comparative genomicists is the well-tuned orchestration of scientific pursuits that unfurl the hidden gems of our DNA encrypted within the molecular signatures of our past and stored in the junkyard of genomes. Hence, exploration of the darker genomic trail of evolution of modern human species is the reflection of our self-exploration that has gone through the cycle of changes in time and space on the Earth.

References

Cheng, K. C. 2008. Skin color in fish and humans: impacts on science and society. Zebrafish. 5(4): 237-42. doi: 10.1089/zeb.2008.0577.

Kolbert, E. (12th March, 2018). There's no scientific basis for race- It's a made-up label.
https://www.nationalgeographic.com/magazine/article/race-genetics-science-africa

Kamberov, Y.G., Wang, S., Tan, J., Gerbault, P., Wark, A., Tan, L., Yang, Y., Li, S., Tang, K., Chen, H., Powell, A., Itan, Y., Fuller, D., Lohmueller, J., Mao, J., Schachar, A., Paymer, M., Hostetter, E., Byrne, E., Burnett, M., McMahon, A.P., Thomas, M. G., Lieberman, D. E., Jin, L., Tabin, C. J., Morgan, B. A., Sabeti, P. C. 2013. Modeling recent human evolution in mice by expression of a selected EDAR variant. Cell 152:691–702. doi:10.1016/j.cell.2013.01.016

Printed in the USA
CPSIA information can be obtained
at www.ICGtesting.com
JSHW011510221024
72173JS00016B/1644/J